NMR

Basic Principles and Progress
Grundlagen und Fortschritte

Volume 5

Editors: P. Diehl E. Fluck R. Kosfeld

With 63 Figures

Springer-Verlag Berlin · Heidelberg · New York 1971

Professor Dr. P. DIEHL
Physikalisches Institut der Universität Basel

Professor Dr. E. FLUCK
Institut für Anorganische Chemie der Universität Stuttgart

Professor Dr. R. KOSFELD
Institut für Physikalische Chemie
der Rhein.-Westf. Technischen Hochschule Aachen

ISBN-13: 978-3-642-65207-3 e-ISBN-13: 978-3-642-65205-9
DOI: 10.1007/978-3-642-65205-9

Softcover reprint of the hardcover 1st edition 1971

NMR

Basic Principles and Progress
Grundlagen und Fortschritte

Volume 5

Editors: P. Diehl E. Fluck R. Kosfeld

With 63 Figures

Springer-Verlag New York · Heidelberg · Berlin 1971

Professor Dr. P. Diehl
Physikalisches Institut der Universität Basel

Professor Dr. E. Fluck
Institut für Anorganische Chemie der Universität Stuttgart

Professor Dr. R. Kosfeld
Institut für Physikalische Chemie
der Rhein.-Westf. Technischen Hochschule Aachen

ISBN-13: 978-3-642-65207-3 e-ISBN-13: 978-3-642-65205-9
DOI: 10.1007/978-3-642-65205-9

Softcover reprint of the hardcover 1st edition 1971

Preface

Nuclear magnetic resonance spectroscopy, which has evolved only within the last 20 years, has become one of the very important tools in chemistry and physics. The literature on its theory and application has grown immensely and a comprehensive and adequate treatment of all branches by one author, or even by several, becomes increasingly difficult.

This series is planned to present articles written by experts working in various fields of nuclear magnetic resonance spectroscopy, and will contain review articles as well as progress reports and original work. Its main aim, however, is to fill a gap, existing in literature, by publishing articles written by specialists, which take the reader from the introductory stage to the latest development in the field.

The editors are grateful to the authors for the time and effort spent in writing the articles, and for their invaluable cooperation.

The Editors

Analysis of NMR Spectra

A Guide for Chemists

R. A. Hoffman†
S. Forsén
Division of Physical Chemistry, Chemical Center, Lund Institute of Technology, Lund, Sweden

B. Gestblom
Institute of Physics, University of Uppsala, Sweden

Contents

I. Principles of NMR Spectroscopy

I.1. The Magnetic Resonance Phenomenon

a) Nuclear Moments

Nuclear magnetic resonance is possible because a number of nuclei possess intrinsic magnetic moments. The existence of an intrinsic magnetic moment is intimately related to the angular momentum properties of the nucleus.

Classically, this relationship may be understood by considering the nucleus as a small but finite rotating charged sphere or ellipsoid, the rotation of which produces a set of infinitesimal current loops whose magnetic fields cooperatively produce a resulting magnetic dipole field (Fig. 1)[1]. The only stable nuclei that possess angular momentum are found to be nuclei with odd mass numbers and the stable even mass isotopes ^{2}H, ^{6}Li, ^{10}B and ^{14}N. Another electromagnetic property of the nuclei which is of some relevance in magnetic resonance applications is the nuclear quadrupole moment which is a measure of the deviations of the nuclear charge distribution from a spherical symmetry.

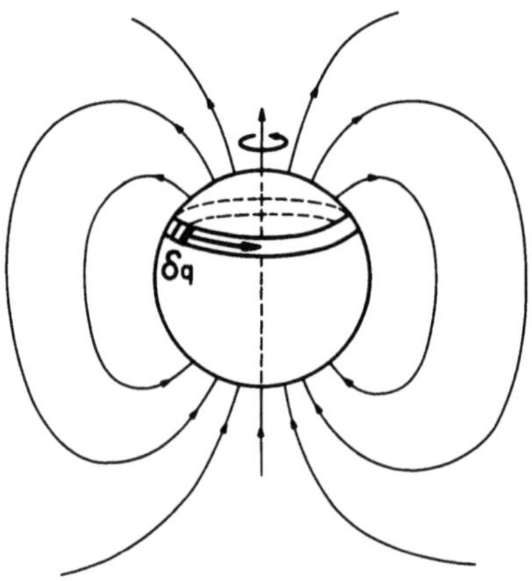

Fig. 1. Classical model of the nucleus as a rotating charged sphere

The angular momentum of a nucleus is quantized, its magnitude being characterized by the nuclear spin quantum number I, which may take on integer or half integer values and is a characteristic number of the isotope considered.

[1] The magnetic moment of a plane current loop is a vector quantity directed perpendicularly to the plane of the current loop and oriented along the direction in which a right hand screw would move if turned in the sense of circulation of the current in the loop. Its magnitude equals the product of the current in the loop and the area enclosed by the current.

The isotopes most commonly employed in magnetic resonance to date, *viz.* ^{1}H, ^{13}C, ^{19}F and ^{31}P all have $I = 1/2$. Such nuclei with $I = 1/2$ cannot possess a nuclear quadrupole moment and this is usually an advantage in high resolution applications where quadrupole interactions tend to produce unwanted line broadening. Among nuclei with I greater than 1/2 one may mention ^{2}H, ^{14}N and ^{10}B as particularly important for NMR applications. Some physical properties of isotopes important in NMR applications are listed in Table 1.

Table 1. *Physical properties of some common nuclei*

Isotope	Spin quantum number I	Relative abundance in nature (%)	Gyromagnetic ratio γ[a]	NMR resonance frequency in MHz at a magnetic field of 10 kilogauss
^{1}H	1/2	99.98	26.7519	42.5759
^{2}H (D)	1	0.016	4.106	6.53566
^{11}B	3/2	81.17	8.853	13.660
^{12}C	0	98.89	—	—
^{13}C	1/2	1.11	6.726	10.705
^{14}N	1	99.64	1.933	3.076
^{15}N	1/2	0.36	− 2.711[b]	4.315
^{16}O	0	99.76	—	—
^{17}O	5/2	0.037	− 3.627[b]	5.772
^{19}F	1/2	100	25.167	40.055
^{27}Al	5/2	100	6.971	11.094
^{31}P	1/2	100	10.829	17.236
^{32}S	0	95.06	—	—
^{33}S	3/2	0.74	2.052	3.266
^{35}Cl	3/2	75.4	2.621	4.172
^{37}Cl	3/2	24.6	2.182	3.472
^{119}Sn	1/2	8.68	− 9.971[b]	15.87
^{199}Hg	1/2	16.92	4.783	7.60

[a] In units of 10^{3} rad $(\text{Gauss})^{-1}$ $(\text{sec})^{-1}$.
[b] The angular momentum vector and the vector of the magnetic dipole are antiparallel.

b) Magnetic Spin States and Energy Levels

When a magnetic dipole is placed in an external magnetic field (H_0) it interacts with the field through a torque tending to align the dipole with the field just as a compass needle tends to orient itself in the earth's magnetic field. The potential energy of a dipole when placed in an external field becomes lowest when the dipole is aligned along the field and highest when the dipole is opposed to the field. A nucleus of spin quantum number I when placed in an external magnetic field H_0 can under stationary conditions only take up $2I + 1$ different inclinations of the angular momentum vector with respect to that field, each inclination being characterized by its own magnetic quantum number *m*. Every one of these stationary states corresponds to a definite energy level and the magnetic resonance signal may be viewed as arising from transitions between these energy levels induced by interactions with a radiation field.

The potential energy of a magnetic moment, $\boldsymbol{\mu}$, in an external magnetic field $\boldsymbol{H}_0$ is given simply by the scalar product:

$$E = -\boldsymbol{\mu} \cdot \boldsymbol{H}_0 . \tag{1}$$

The magnetic moment vector, $\boldsymbol{\mu}$, is parallel with the spin angular momentum vector, $\boldsymbol{I}\hbar$, and is given by

$$\boldsymbol{\mu} = \gamma \boldsymbol{I} \hbar \tag{2}$$

where the constant of proportionality, γ, is called the gyromagnetic ratio of the nucleus and $\hbar$ stands for Planck's constant divided by 2π and is an entity of the dimension of an angular momentum.

At this point it becomes convenient to introduce a frame of reference. The Cartesian coordinate systems employed in NMR are always chosen so as to have their $x-y$ plane perpendicular to the magnetic field H_0. We shall use such a coordinate system, with x and y axes fixed in relationship to the spectrometer probe ("the laboratory frame") and oriented so as to bring the H_0 field along the *negative* z direction. Inserting Eq. (2) into Eq. (1), we obtain in this frame of reference the energy expression:

$$E = +\gamma \hbar H_0 I_z \tag{3}$$

where I_z is the z component of the dimensionless vector $\boldsymbol{I}$. (Note that Eq. (3) would have retained the minus sign of Eq. (1) if the positive z direction had been chosen along H_0.)

The component I_z assumes a definite value m in a stationary state of the system (I_z is a "constant of the motion") and Eq. (3) may be rewritten in terms of the *magnetic quantum number m*:

$$E = \gamma \hbar H_0 m . \tag{4}$$

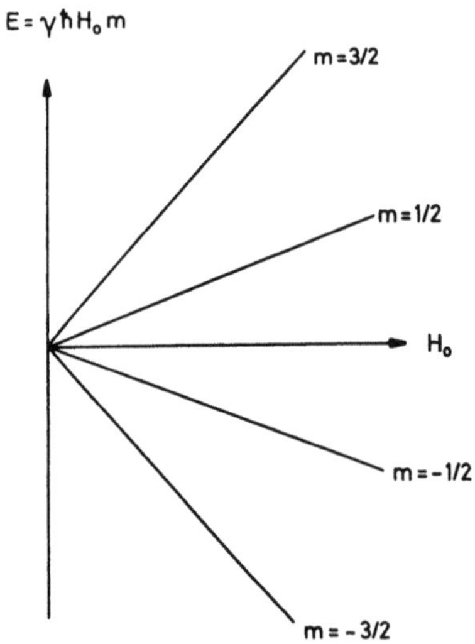

Fig. 2. The spin energy of a nucleus as a function of the magnetic field H_0. The nucleus has the spin quantum number $I = 3/2$

The magnetic quantum number m of Eq. (4) can assume any one of the $2I+1$ values

$$\begin{aligned} m &= I \\ m &= I-1 \\ &\vdots \\ m &= -I\,. \end{aligned}$$

For example, if $I=1/2$, m can assume the values of $+1/2$ or $-1/2$, and if $I=1$ the possible values of m become $+1$, 0 and -1. It follows that the energy of a magnetic nucleus under the influence of the magnetic field H_0 may assume any one of $2I+1$ equidistant values, each energy level being characterized by its magnetic quantum number m (Fig. 2).

c) The Magnetic Resonance Condition

The selection rule governing the possible transitions in an NMR experiment is

$$\Delta m = \pm 1 \tag{5}$$

and the Bohr frequency condition $h\nu = \Delta E$ gives the magnetic resonance equation

$$h\nu = \gamma \hbar H_0\,. \tag{6}$$

The magnitude of the level separations is such that the frequency ν of the radiation field quanta (photons) $h\nu$ falls in the radiofrequency region of the electromagnetic spectrum when the applied field H_0 is of a magnitude conveniently generated in the laboratory. In this region the source of radiation will be a radio frequency coil fed by a radio frequency transmitter and this coil is so oriented that it produces an oscillating magnetic field H_1 polarized perpendicular to the steady field H_0.

d) The Larmor Precession

In radiofrequency spectroscopy the radiation field in the sample volume is almost perfectly coherent, both temporally and spatially. Before the advent of lasers, spectroscopy with such completely coherent radiation sources was virtually unknown in the optical domain and the traditional mode of presenting the elementary theory of spectra does not lend itself well to description of coherence phenomena. In order to take these into account in NMR spectroscopy, one therefore has recourse to a classical model of the nucleus as a spinning charged top. When a spinning top experiences a torque tending to turn over the spinning axis, the resulting motion becomes a precessional motion, the spin axis moving along the surface of a circular cone. A mechanical analogue of a spinning magnetic nucleus is a heavy symmetrical top placed on horizontal support. Under the torque produced by the opposing forces of the gravitational pull at the center of gravity and the upward reaction force from the support, the spinning top does not turn over but precesses around the vertical field lines of gravity (Fig. 3). The precessional frequency, ν, of a spin magnetic moment $\boldsymbol{\mu}=\gamma \boldsymbol{I}\hbar$ in an external magnetic field $\boldsymbol{H}_0$ is given by the Larmor equation

$$\nu=\gamma H_0/2\pi \tag{7}$$

which classical equation is exactly equivalent to the quantum mechanical relation Eq. (6). A radiofrequency field oscillating in a plane perpendicular to H_0 at a frequency given by Eq. (7) will be in resonance with the precessional motion of the spinning nuclei, thus making possible an energy exchange between the spin system and the radiofrequency field.

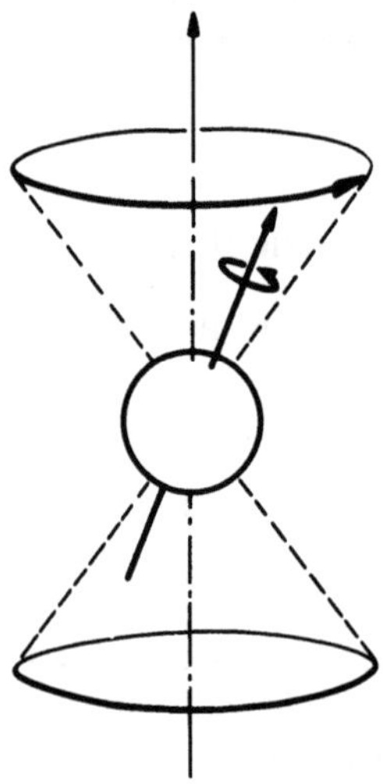

Fig. 3. Precession of a spinning top around the vertical field lines

e) Experimental Aspects

The occurence of a magnetic resonance may be detected either as a change in impedance of a single coil probe, the energy dissipation corresponding to the magnetic resonance absorption being experienced as an extra resistive contribution to the coil impedance, or in a two-coil probe as an induced voltage in the receiver coil produced by the precessing magnetic moment of the ensemble of nuclei in the sample.

In order to scan the spectrum one may vary the radiofrequency of the transmitter at a given polarizing magnetic field H_0 so as to fulfill the resonance condition (Eqs. (6) or (7)). This *frequency sweep* method has many advantages over the traditional *field sweep* method in which one varies the H_0 field so as to bring the energy level separation into resonance with a constant radiofrequency. It will be noted that this latter method of scanning a spectrum by slowly varying the level separations so as to bring them into resonance with a monochromatic radiation source cannot be realized in optical molecular spectroscopy where the energy level separations are intrinsic molecular properties not under experimental control.

I.2. Chemical Shifts

a) The Screening Constant σ

The resonance conditions, Eqs. (6) and (7), were derived by considering only the interaction between nuclear magnetic moments and the magnetic field H_0 of the laboratory magnet. They are easily generalized, however, to include the screening effect of the electrons in the molecule. In an actual sample, the externally applied magnetic field H_0 induces currents in the electron clouds of the molecules, and these currents in turn produce their own secondary magnetic fields at the sites of the magnetic nuclei, thus shifting the resonance away from that of a bare nucleus (the *chemical* shift). This induced secondary field is proportional in magnitude to the externally applied H_0 field and in *isotropic* environments its direction is of necessity along the H_0 field.

Thus the effective magnetic field $\boldsymbol{H}$ at the nucleus is given by

$$\boldsymbol{H} = \boldsymbol{H}_0 - \sigma \boldsymbol{H}_0 \tag{8}$$

where σ is a dimensionless constant, called the (isotropic) *screening constant* of the nucleus in the environment considered.

Isotropic surroundings are effectively produced in fluid media (liquids, solutions and gases) and these are also the media where high-resolution NMR is normally ewployed (cf. Section II.2, Example 1). For hydrogen nuclei in diamagnetic samples the screening constant σ varies over a range of some 10 to 20 ppm. Screening constants of heavier nuclei tend to be a few orders of magnitude larger because of the greater ease of inducing currents in their electron clouds. Insofar as the sizes of the currents induced in a charge distribution

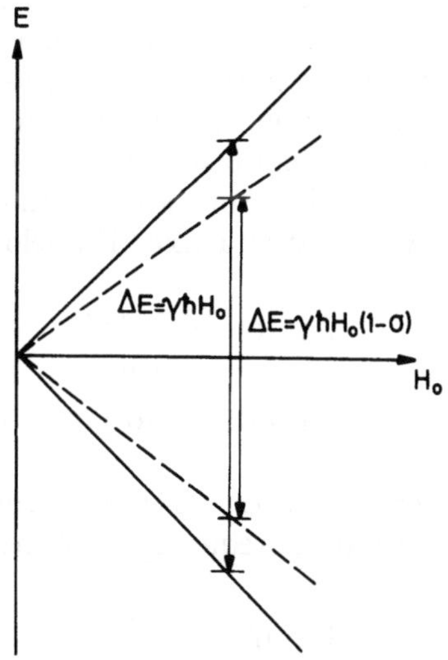

Fig. 4. The spin energy of a nucleus (I = 1/2) as a function of the magnetic field H_0. The full lines show the energy levels of the bare nucleus, while the broken lines show the energy levels when the chemical shift (screening constant σ) is taken into account. The screening effect has been exaggerated in the figure

depend on the amount of charge present, one may expect a simple correlation between screening constants (chemical shifts) and charge densities, but this correlation must not be taken for granted.

The magnetic energy of a screened nucleus in an externally applied field H_0 is obtained by a slight generalization of Eq. (4) and is given by

$$E = \gamma \hbar H_0 (1 - \sigma)\, m \tag{9}$$

with the H_0 of Eq. (4) replaced by the effective magnetic field H of Eq. (8). The energy expression (9) when combined with the selection rule of Eq. (5) and the Bohr frequency condition $h\nu = \Delta E$ yields the resonance condition

$$\nu = \frac{\gamma}{2\pi} H_0 (1 - \sigma) \tag{10}$$

for a nucleus of gyromagnetic ratio γ in an environment described by a screening constant σ and subject to an applied magnetic field H_0.

The frequency ν defined by Eq. (10) will henceforth be called the *Larmor frequency* of the nucleus in question, whether or not additional interactions displace (or split) the resonance frequency away from that given by Eq. (10).

The effect of a chemical shift on the energy levels and the transition frequency of a nucleus is illustrated in Fig. 4.

b) Chemical Shift Scales (σ and τ)

Since the resonance frequency of the bare nucleus is seldom known, screening constants and chemical shifts are usually related to some standard reference. In proton magnetic resonance spectroscopy the generally accepted standard reference is tetramethylsilane $(CH_3)_4Si$ (TMS). The screening constant of TMS is larger than that of most other protons and, if the spectrum is recorded by sweeping the magnetic field while using a constant frequency of the stimulating radiofrequency field, the TMS signal will appear upfield from most of the other absorption lines.

Quantitatively, the chemical shift δ relative to the reference (TMS) of some proton X in a sample is defined in terms of the Larmor frequencies ν_X and ν_{TMS} as given by Eq. (10) and measured at a *fixed* H_0 field. Thus δ is defined by the expression

$$\delta = 10^6 \frac{\nu_X - \nu_{TMS}}{\nu_{TMS}} \tag{11}$$

where the factor 10^6 has been introduced to give δ in the convenient units of parts per million (ppm).

In the right-hand member of Eq. (11) the denominator ν_{TMS} is usually replaced by the nominal operating frequency ν_0 of the spectrometer to yield the simple equation,

$$\delta = 10^6 \frac{\nu_X - \nu_{TMS}}{\nu_0}\,. \tag{12}$$

The difference $(\nu_X - \nu_{TMS})$ can be read off directly on a precalibrated recorder chart or obtained more accurately by modulation side-band techniques so that the δ values of Eq. (12) are easily evaluated.

It should be pointed out at this stage that the definition of Eq. (11) gives a unique chemical shift value for a given proton in a given sample, whereas the rather loose definition of Eq. (12) does not, but to the three or four significant figures usually quoted for chemical shifts this difference is unobservable.

By introduction of the definitions of Eq. (10) for ν_X and ν_{TMS} in Eq. (11), we find that δ is related to the screening constants σ_X and σ_{TMS} through the equation

$$\delta = 10^6 \frac{\sigma_{TMS} - \sigma_X}{1 - \sigma_{TMS}} \tag{13}$$

and since σ_{TMS} is only of the order of 10^{-5}, it follows that δ is practically equal to the difference in screening constants $(\sigma_{TMS} - \sigma_X)$ as expressed in ppm.

The definition of Eq. (11) is directly applicable to a frequency sweep spectrum. If the spectrum is instead recorded by sweeping the magnetic field at a constant spectrometer frequency ν_0, the resonant values of the magnetic field, *viz.* H_0 for the sample and $(H_0 + \Delta H_0)$ for the TMS reference, are related by the equations

$$\nu_0 = \frac{\gamma}{2\pi} H_0(1 - \sigma_X) = \frac{\gamma}{2\pi} (H_0 + \Delta H_0)(1 - \sigma_{TMS}). \tag{14}$$

From the last two members of Eq. (14) one can derive the expression

$$\frac{\Delta H_0}{H_0} = \frac{\sigma_{TMS} - \sigma_X}{1 - \sigma_{TMS}} \tag{15}$$

and it follows from Eqs. (13) and (15), that δ may be defined as

$$\delta = 10^6 \frac{\Delta H_0}{H_0}. \tag{16}$$

In actual practice, the magnetic fields H_0 and ΔH_0 are never measured directly, and the δ values quoted by experimentalists are always those defined by Eq. (12) with $(\nu_X - \nu_{TMS})$ obtained by precalibration or by modulation side-band techniques. In terms of the sweep variable then, the quantity $(\nu_X - \nu_{TMS})$ represents *either* the difference between the frequencies used to excite the sample and reference resonances at a fixed H_0 field, *or* the frequency equivalent $\Delta \nu$, given by

$$\Delta \nu = \Delta H_0 \gamma / 2\pi \tag{17}$$

of the difference ΔH_0 between the magnetic fields $(H_0 + \Delta H_0)$ and H_0 used to excite the reference and sample resonances at a fixed spectrometer frequency ν_0.

In proton magnetic resonance the sweep variable (H_0 or ν_0) normally varies by less than 10 or 20 ppm over the high-resolution spectrum. If one abstains from double-irradiation experiments, one then obtains field-sweep spectra that are indistinguishable from the corresponding frequency-sweep spectra as long as relative accuracies of 10^{-5} are not attainable. It will be noted, however, that the correspondence requires opposite directions of the two sweeps. Thus a positive δ requires according to Eqs. (14)–(16) that the X-resonance is *downfield* from TMS and, according to Eq. (12), that the sample resonance frequency ν_X

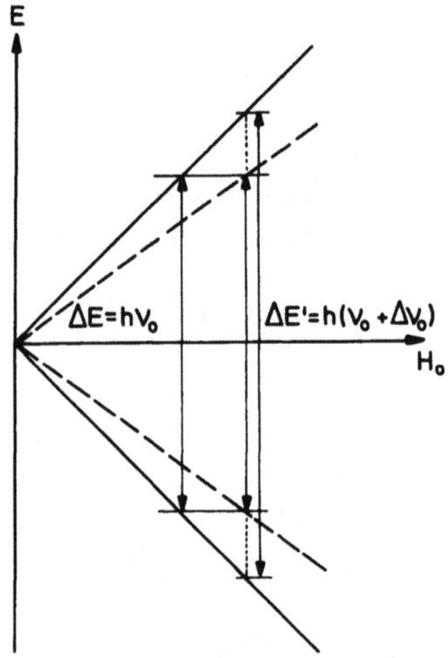

Fig. 5. Energy levels and transitions of two nuclei (I = 1/2) which are chemically shifted relative to each other. It is seen that at constant field H_0 the more screened nucleus will be observed at a lower radiofrequency than that of the less screened nucleus. If the radiofrequency is fixed, i.e. $\Delta E = h\nu_0$ is constant, the more screened nucleus will be observed at a higher H_0 field than the less screened nucleus

will be *higher* than that of TMS when compared at a given H_0 field. This inversion of relative magnitudes is illustrated graphically in Fig. 5.

The universally accepted mode of displaying NMR spectra is a left-to-right arrangement in which H_0 in a field sweep spectrum increases from left to right and, consequently, ν in a frequency sweep spectrum decreases from left to right.

The frequency sweep mode of operation is the one which has a counterpart in other fields of spectroscopy and, for this reason, it is the conceptually simpler method. The sense of *increasing* frequency (and hence increasing transition energy) in the frequency-sweep mode corresponds to a *downfield* sweep in the field-sweep mode. In our opinion this represents a very strong argument in favour of a sign convention in which the frequency scale (and the chemical shift) *increases downfield* from TMS. This is the convention used in the definition of δ and is the one recommended for proton magnetic resonance by the American Society for Testing and Materials (ASTM, Manual of February 1966, sponsored by committee E-13) but is contrary to that of the τ-scale, introduced by TIERS [*1*] and recommended by the Chemical Society of London. The parameter τ is defined as

$$\tau = 10 - \delta \,. \tag{18}$$

I.3. Spin Coupling Constants

So far we have only considered the interaction between the magnetic moment of a nucleus and a stationary magnetic field H_0 or H, produced by the laboratory magnet and modified by induced electronic currents. In actual spin systems, however, interactions between different nuclear spins in a molecule produce additional structure in the NMR spectrum. In isotropic fluids this interaction is not caused by direct "through space" interactions of the individual magnetic dipoles because this interaction averages out to zero through the rapid random reorientations of the internuclear axes. There remains, however, a small isotropic interaction between nuclear spins mediated by the electron-nuclear and electron-electron interactions in the molecule. The energy of this interaction is proportional to the scalar product of the two spin vectors involved and may be written

$$E = hJ_{AB}\boldsymbol{I}(A) \cdot \boldsymbol{I}(B) \tag{19}$$

where $\boldsymbol{I}(A)$ and $\boldsymbol{I}(B)$ are the spin vectors of the coupled nuclei A and B and where the constant J_{AB} has the dimension of frequency and is called the spin-coupling constant between A and B.

Example 1: Spin-spin Splitting in the HD Molecule

The mechanism of spin copling and its effects on the high-resolution spectrum will now be briefly discussed with reference to spin-spin splitting in the HD NMR spectrum at $H_0 = 14.100$ gauss, taken as an example.

The orienting effect of the magnetic field H_0 in this case is very much larger than that of the spin-spin coupling and both the proton and the deuteron will be individually characterized by their magnetic quantum numbers, m_H and m_D, in a stationary state. Since $I(H) = 1/2$ and $I(D) = 1$, the possible values of m_H are $+1/2$ and $-1/2$, while the possible values of m_D are $+1$, 0 and -1. The magnetic energy of the spin system is very small compared to its thermal energy and every magnetic quantum state will be occupied with (almost) the same probability (the Boltzmann factors of the different levels are very nearly equal).

Now consider a molecule in which $m_H = -1/2$, *i.e.* the proton is aligned with the external field H_0. Because of magnetic interactions between the proton spin and the 1s electrons of the molecule, the electron spin when close to the proton becomes slightly polarized and tends to oppose the external field. The spins of the two electrons forming the bond are, however, strongly correlated and a polarization of an electron close to the proton favours an electron density of opposite spin close to the deuteron. The magnetic interaction between this electron spin density around the deuteron and the deuteron magnetic moment in turn results in a spin coupling *contribution* to the total energy of the spin system which is highest when the deuteron and the proton have parallel spins and lowest when their spins are opposed. This state of affairs is described by a positive value of the spin-coupling constant[2].

The proton magnetic resonance spectrum may be considered to arise from transitions between the $m_H = +1/2$ and $m_H = -1/2$ states of the proton in three different sets of molecules, *viz.* those having $m_D = +1$, $m_D = 0$ and $m_D = -1$.

[2] From the foregoing it should be obvious that not only the magnitude but also the sign of the spin-coupling constant is an intrinsic molecular parameter determined by the mechanism of spin polarization transfer in the molecule. Unfortunately, however, the *absolute* signs of spin-coupling constants cannot be determined from high resolution NMR spectra except in very special cases. However, the *relative* signs of different spin couplings in complex spin systems can generally be determined either by a detailed analysis of the spectrum or by double-irradiation techniques.

Every member of this set of molecules produces its own proton resonance line, the frequency separation between the separate lines being equal to J_{HD}. Since every member of the set also has approximately the same population, the three proton absorption lines will be of equal intensity, yielding a 1:1:1 triplet.

The observed transition frequencies may be calculated analytically from an expression for the total magnetic energy of the spin system by use of the selection rules,

$$\Delta m_H = +1\,; \quad \Delta m_D = 0\,, \tag{20}$$

for the proton absorption spectrum, and

$$\Delta m_D = +1\,; \quad \Delta m_H = 0 \tag{21}$$

for the deuteron absorption spectrum. These selection rules are modifications of Eq. (5) and express the facts that the non-resonant nucleus does not change its quantum state, and that absorption rather than emission ($\Delta m = -1$) is considered.

The total magnetic energy of the nuclear spins in the H–D molecule is composed additively by three contributions, *viz.*

a) the interaction energy of the proton individually with the polarizing field (the proton Zeeman energy),

b) the corresponding energy of the deuteron, and finally,

c) the mutual spin-spin coupling energy.

The nuclear Zeeman energy of a) and b) together, may be written

$$E = h(\nu_H m_H + \nu_D m_D) \tag{22}$$

as follows from Eqs. (9) and (10) if ν_H and ν_D are taken as the proton and deuteron Larmor frequencies respectively. The spin coupling energy c) is given by Eq. (19) but this expression may be simplified in the present case for the following reason.

The spin-spin coupling interaction represents a very weak orienting effect, and at $H_0 = 14.100$ gauss the spin vectors $\boldsymbol{I}$(H) and $\boldsymbol{I}$(D) can have no components that are transverse to $\boldsymbol{H}_0$ in a stationary state. Consequently, the spin vectors $\boldsymbol{I}$(H) and $\boldsymbol{I}$(D) are represented effectively by their z-components I_z(H) and I_z(D) and the scalar product $\boldsymbol{I}(H)\cdot\boldsymbol{I}(D)$ of Eq. (19), which in general decomposes according to

$$\boldsymbol{I}(A)\cdot\boldsymbol{I}(B) = I_x(A)\,I_x(B) + I_y(A)\,I_y(B) + I_z(A)\,I_z(B) \tag{23}$$

reduces to the single term $I_z(H)\,I_z(D)$. In a stationary state, the product $I_z(H)\,I_z(D)$ may in turn be rewritten in terms of the magnetic quantum numbers as $m_H m_D$. Thus the total nuclear magnetic energy of the H–D molecule may be expressed in terms of the magnetic quantum numbers m_H and m_D and is given by

$$E = E(m_H, m_D) = h[\nu_H m_H + \nu_D m_D + J_{HD} m_H m_D]\,. \tag{24}$$

The proton resonance frequencies are obtained by use of Eq. (20) and Bohr's frequency condition $h\nu = \Delta E$ yielding the frequencies

$$\nu_{\text{proton}} = \frac{E(m_H + 1, m_D) - E(m_H, m_D)}{h}\,. \tag{25}$$

Insertion of Eq. (24) into Eq. (25) gives the expression

$$\nu_{\text{proton}} = \nu_{\text{H}} + J_{\text{HD}}\, m_{\text{D}}\,. \tag{26}$$

Eq. (25) gives one proton transition frequency for every value of m_D, *i.e.* three in all. The centre peak, corresponding to $m_D = 0$, occurs at the same frequency as would have been obtained if no spin coupling had been present. The resulting spectrum is depicted in Fig. 6b. At the field of 14.100 gauss considered, the proton resonance spectrum will be found at a frequency of 60.00 MHz (cf. Table 1). The deuteron resonance spectrum is obtained similarly by combination of Eqs. (21) and (24) to yield

$$\nu_{\text{deuteron}} = \nu_{\text{D}} + J_{\text{HD}}\, m_{\text{H}} \tag{27}$$

and consists of a 1:1 doublet with a splitting of J_{HD}, centered around the frequency ν_D that would have been obtained in the absence of spin-coupling effects (Fig. 6a). At 14.100 gauss, this frequency will of the order of 9.21 MHz.

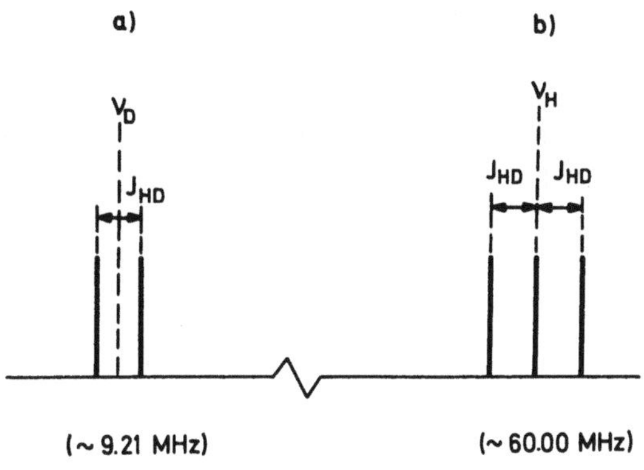

Fig. 6. Schematic NMR spectrum of HD illustrating the effects of spin-spin coupling. a The deuteron resonance spectrum and b the proton resonance spectrum

In larger systems the number of NMR parameters (chemical shifts and spin-coupling constants) generally increases rapidly with the number of spins and the spectra increase in complexity accordingly.

I.4. Intensities

The absolute intensity of an NMR signal is a somewhat complex function of sample parameters, instrument parameters and operating conditions. On the other hand, the relative intensities of different bands in a typical NMR spectrum do often follow simple rules. Thus, if the nuclei (protons) of one functional group have a chemical shift such that the total absorption band of this group is well separated from other absorption bands in the spectrum, then the *integrated intensity* of this band is proportional to the number of nuclei (hydrogens) of that functional group.

Example 1: Relative Intensities in Diethyl Ether

The proton magnetic resonance spectrum of diethyl ether consists of two quite distinct absorption bands at $\delta = 5.9$ ppm and $\delta = 8.8$ ppm and the widths of the bands are less than 0.4 ppm. The integrated intensities of the two bands are in the ratio 2:3.

Example 2: Relative Intensities in Indene

An example of a more complex spectrum that may be decomposed into different bands assignable to functional groups is that of indene (I) at 60 MHz,

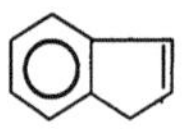

(I)

which shows four groups of multiplets at $\delta = 7.10$, $\delta = 6.65$, $\delta = 5.97$ and $\delta = 2.99$ ppm. The relative intensities of these bands are very close to the integer ratios 4:1:1:2. These numbers represent the number of protons in the six membered ring (4), in the 3-position (1), in the 2-position (1) and in the 1-position (2), respectively.

These simple intensity relationships are basic to most analytical applications of NMR. The relative intensities of the different lines within a multiplet and in spectra where the chemical shift differences are not large compared with the spin-spin splittings, will be dealt with in part IV.

II. Some Basic Concepts

II.1. Groups of Equivalent Spins

The structure of a high-resolution NMR spectrum depends among other things on whether the magnetic nuclei of the molecule may be divided into groups of equivalent spins, or not. It is important to distinguish between different types of equivalence. The three principal classes of equivalence are:

i. chemical shift equivalence,
ii. symmetrical equivalence,
iii. magnetic equivalence.

In this enumeration magnetic equivalence implies the highest degree of indistinguishability (from an NMR point of view) of the nuclei concerned and chemical shift equivalence the lowest. The concepts of equivalent nuclei are applicable only to nuclei of the same isotopic species.

a) Definitions of Equivalence Groupings

i. Chemical shift equivalence implies only that the nuclei in question have the same chemical shift.

ii. Symmetrical equivalence is encountered in molecules displaying some kind of symmetry and implies that a symmetry operation on the molecule, will exchange the positions of symmetrically equivalent nuclei. Symmetrically equivalent nuclei are of necessity chemical shift equivalents.

The applicability of this definition is straightforward only in rigid molecules. When symmetrical equivalence results from motional averaging (or accidentally) an alternative formulation is more helpful.

iii. Symmetrical Equivalence: Alternative Formulation. Symmetrical equivalence exists in a spin system if a permutation of the spins can be found that leaves the values of all the NMR parameters (chemical shifts and spin couplings) invariant. This definition should be interpreted strictly pragmatically and the NMR parameters to be considered are the effective values resulting from all the interactions including motional averaging over different conformations in a non-rigid molecule. If every permutation of the spins within a group of nuclei leaves the effective NMR parameters of the spin system invariant, then the spins within that group are in addition magnetically equivalent. However, if invariance of the NMR parameters requires that a permutation of the spins in one group be compensated by some permutation of spins in another group, then the nuclei are only symmetrically equivalent. In a rigid molecule only those permutations should be considered that result from a symmetry operation on the molecule: an example of a prohibited permutation is the exchange of only two protons in one of the ^{13}C-ethylenes (Example 3 below).

iv. Magnetic equivalence of a group of nuclei A1, A2, A3 ... An, requires that all the nuclei of that group have the same chemical shift and, in addition, that if any nucleus, say A1, of that group is spin-coupled with some nucleus (B) *outside* the group, then the remaining nuclei A2, A3 ... An of the group must be spin-coupled to the same extent with nucleus B[3].

For example, if the molecule has two magnetic nuclei B and X in addition to the group A_n of magnetically equivalent nuclei, then the conditions for magnetic equivalence of the n "A" spins are

$$\sigma_{A1} = \sigma_{A2} = \sigma_{A3} = \ldots \sigma_{An} \tag{1}$$

and

$$J_{A1,B} = J_{A2,B} = J_{A3,B} = \ldots J_{An,B} = J_{AB} \tag{2}$$

as well as

$$J_{A1,X} = J_{A2,X} = J_{A3,X} = \ldots J_{An,X} = J_{AX} \tag{3}$$

but equality of J_{AB} and J_{AX} is not required.

Magnetic equivalence implies that the equivalent nuclei are indistinguishable in all their NMR interactions and can be permuted at will without changing the physical state of the spin system. The nuclei of such a group behave as a single entity ("composite particle") in their interactions with the *outside* world (i.e. in their interactions with macroscopic magnetic fields) and *mutual spin couplings within a group of magnetically equivalent nuclei produce no observable effects in the NMR spectrum.* For convenience, these unobservable spin couplings are therefore arbitrarily set equal to zero when calculating NMR spectra of molecules contàining magnetically equivalent nuclei.

The above definitions will now be illustrated by a few examples and their usefulness will be discussed in connection with the examples.

Example 1: Chemical Shift Equivalence

The two protons in the 2- and 5-positions of 3-d-thiophene (I)

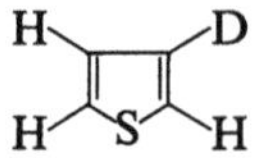

(I)

will, neglecting the small deuterium isotope effects on the proton chemical shifts, display chemical shift equivalence and chemical shift equivalence only. Because of strong spin-coupling effects, the chemical shift equivalence produces no simplification of the spectrum in this case [2].

Unless a molecule displays some kind of symmetry, the occurrence of chemical shift equivalence is always more or less accidental. If the nuclei of a chemical shift equivalent group are not spin-coupled to any nucleus outside that group, then the nuclei are by definition magnetically equivalent. More generally it may

[3] This interpretation of the concept of magnetic equivalence is the generally accepted one, even though (at least) one author has triéd to get away with the confusing use of this term for chemical shift equivalence.

be said that chemical shift equivalence (*i*) produces no simplification of the spectrum unless this equivalence is extended to one of the two higher forms of equivalence (*ii*) or (*iii*).

Example 2: Magnetic Equivalence in Normal Ethylene

The protons of ethylene (II) are chemical shift

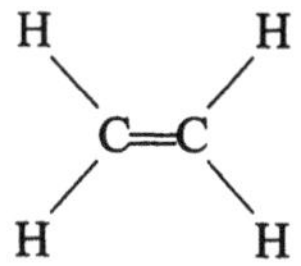

(II)

equivalent. In ethylene, the chemical shift equivalence follows from symmetry and the protons may also be classified as being *symmetrically equivalent.* In ethylene which has not been isotopically enriched in ^{13}C, 97.8% of the molecules will contain only the isotope ^{12}C, which has no magnetic moment ($I=0$) and cannot be spin-coupled to the protons. In ordinary ethylene the protons will therefore also be *magnetically equivalent* in the vast majority of molecules. Consequently these molecules contain only magnetic nuclei that are all magnetically equivalent, and – since spin-coupling among magnetically equivalent nuclei is unobservable – their magnetic resonance spectrum will contain only one sharp absorption line, centered at the ethylene proton Larmor frequency. The predominant feature in the proton magnetic resonance spectrum will be this structureless peak assigned to the 97.8% abundant species containing only ^{12}C nuclei in the carbon sceleton.

Example 3: Magnetic Non-equivalence in ^{13}C-Ethylenes

In 2.2% of the ethylene molecules one of the carbon nuclei will be the isotope ^{13}C.

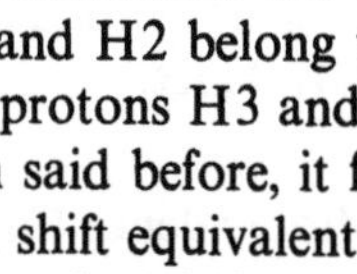

(III)

The asymetric isotopic substitution reduces the symmetry of the molecule but the directly bonded protons H1 and H2 belong to one group of symmetrically equivalent nuclei and the vicinal protons H3 and H4 belong to a second group of this type. From what has been said before, it follows that these proton pairs also form two groups of chemical shift equivalent nuclei. Disregarding the small ^{13}C isotope effect on the proton chemical shifts ($\pm 15 \cdot 10^{-4}$ ppm), all four protons in (III) will form one single group of chemical shift equivalent nuclei. Even so, they would not be magnetically equivalent since the ^{13}C nucleus has a spin of $I=1/2$ and displays widely unequal spin-couplings to the directly bonded ($J=156.4$ Hz) and the vicinal ($J=2.5$ Hz) protons. It is interesting to note that the magnetic non-equivalence introduced by the ^{13}C nucleus makes the proton spectrum depend on all the proton spin-couplings, whereas in the spectrum of the more abundant ethylene molecules containing only ^{12}C nuclei it depended on none. The procedure to extract the values of the spin-coupling constant in a spectrum like that of (III) will be outlined in Section IV.9.

Normal ethylene will also contain some 0.01% of molecules containing two ^{13}C nuclei (IV). In this

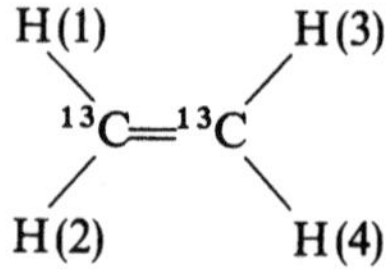

(IV)

molecule all four protons are strictly symmetrically equivalent. However, no group of magnetically equivalent nuclei is to be found in this molecule! This may be seen in the following way: protons H1 and H2 are not magnetically equivalent with protons H3 and H4 since these two pairs of nuclei are unequally coupled to a particular ^{13}C nucleus. Nor will protons H1 and H2 then form a group of magnetically equivalent nuclei because the rule that all the nuclei in a magnetically equivalent group must be equally spin-coupled to a nucleus outside that group is violated by the inequality of $J_{H1,H4}$ (19.0 Hz) and $J_{H2,H4}$ (11.7 Hz).

Example 4: Magnetic Equivalence through Motional Averaging

Magnetic equivalence often occurs as a result of rotational averaging as exemplified by the molecule 1,1,1-trifluoroethane (V)

$$CF_3{-}CH_3$$

(V)

containing only the most abundant isotopes. If in this molecule one particular rotational conformer were frozen in, the three protons and the fluorine nuclei would not form groups of magnetically equivalent nuclei since the vicinal H–F spin couplings depends on the F–C–C–H dihedral angle [*3*]. At room temperature, however, rapid internal rotation about the C–C bond produces averaged H–F spin couplings that are the same for all conceivable H–F pairs.

Magnetic equivalence caused by motional averaging is commonly encountered in molecules containing aliphatic groups. However, there are numerous cases in which rapid internal rotation does not produce magnetic equivalence or even symmetrical or chemical shift equivalence.

Example 5: Chemical Shift Non-Equivalence Caused by an Asymmetrical Center

When a methylene group is attached to an asymmetric grouping, the chemical shifts of the two methylene protons need not be identical, as will now be demonstrated for a 1,1,2-trisubstituted ethane. The factors primarily responsible for the chemical shift non-equivalence depend on the temperature range considered. At very low tewperatures the rotation around the C–C bond is virtually frozen in. In any particular rotational isomer the environments of the two methylene protons differ, so that even if only one conformer is stable one must *a priori* expect the two protons to display unequal chemical shifts. At somewhat higher temperatures the different rotational isomers are rapidly interconverted but the residence times of different conformers would still be quite different. In this temperature range the chemical shift difference of the methylene protons follows frow simple statistical averaging (Fig. 7). If the probabilities of the three

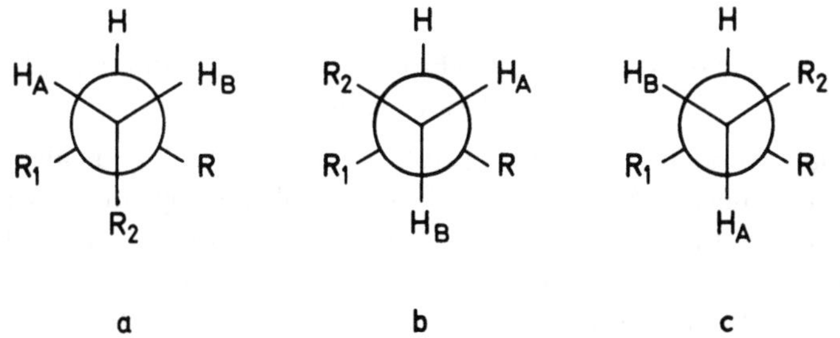

Fig. 7. The three rotational isomers of a 1,1,2-trisubstituted ethane

staggered conformations (assumed dominant) are denoted by P_a, P_b and P_c, and the screening constant of protons A and B are designated by σ_A^a and σ_B^a in conformer *a* etc., the resulting averaged screening constants σ_A and σ_B are given by

$$\sigma_A = P_a \cdot \sigma_A^a + P_b \cdot \sigma_A^b + P_c \cdot \sigma_A^c , \tag{4}$$

$$\sigma_B = P_a \cdot \sigma_B^a + P_b \cdot \sigma_B^b + P_c \cdot \sigma_B^c . \tag{5}$$

Since the screening constants of A and B are unequal in any given conformer, equality of the averaged values is exceedingly improbable when the probabilities P_a, P_b and P_c are different. At still higher temperatures, when the probabilities P_a, P_b and P_c are all very close to 1/3, this explanation for the chemical shift difference is no longer valid and must be replaced by reference to the fact that dissimilarities in the *overall* environment tend to produce unequal screening constants of protons A and B, even in those conformer pairs that place A and B in equivalent positions with respect to the substituents on the vicinal carbon atom, thus, in general

$$\sigma_A^a \neq \sigma_B^c ,$$
$$\sigma_A^b \neq \sigma_B^a ,$$
$$\sigma_A^c \neq \sigma_B^b .$$

The chemical shift non-equivalence does not require the molecule as a whole to be asymmetric and resolvable into optical antipodes, nor does it require the asymmetric center to be found on the nearest neighbour atom. The arguments presented above may be used to demonstrate that non-equivalent chemical shifts of methylene protons may arise in any molecule containing a carbon atom with three non-equivalent bonds, and many examples have been reported in the literature. For example, the geminal methylene protons in anyone of the two ethyl groups of acetaldehyde diethylacetal (VI)

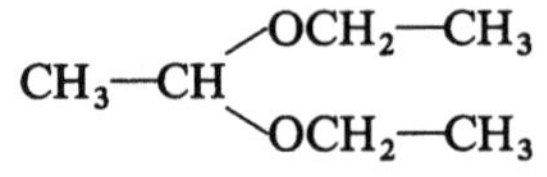

(VI)

have been found to have unequal chemical shifts (cf. the review by SIDALL and STEWART [4]).

Example 6: Magnetic Non-Equivalence in Unsymmetrically Substituted Ethanes

Motional averaging may also result in symmetrical equivalence without magnetic equivalence. An example of this is offered by an unsymmetrically 1,2-disubstituted ethane (XCH_2–CH_2Y). At low temperatures, when the rotation around the C–C bond is slow enough to produce separate spectra for each rotational isomer, the geminal proton pairs of the *trans* conformer form pairs of symmetrically equivalent nuclei (Fig. 8) but the protons of other conformers

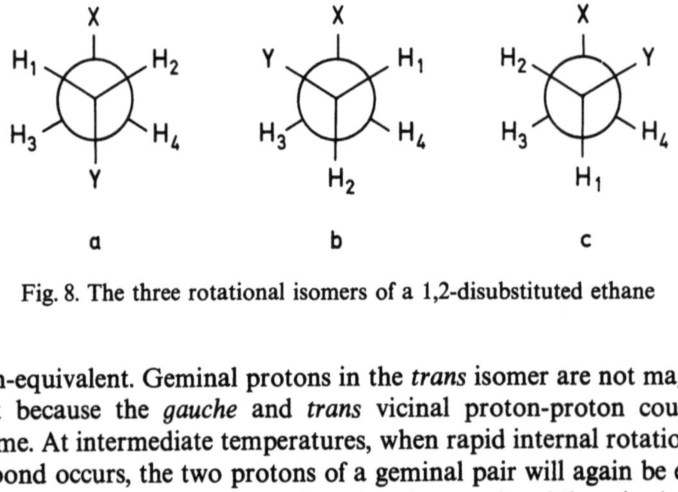

Fig. 8. The three rotational isomers of a 1,2-disubstituted ethane

are all non-equivalent. Geminal protons in the *trans* isomer are not magnetically equivalent because the *gauche* and *trans* vicinal proton-proton coupling are not the same. At intermediate temperatures, when rapid internal rotation around the C–C bond occurs, the two protons of a geminal pair will again be effectively symmetrically equivalent as a result of motional averaging. (The criterion applied in the present case is the fulfillment of the conditions stated in the alternative definition of symmetrical equivalence in terms of spin permutations.) A permutation in which H 1 is exchanged with H 2 and H 3 with H 4 leaves the motionally averaged value of the NMR parameters invariant since the Eqs. (6) to (9)

$$\sigma_1 = \sigma_2 , \tag{6}$$

$$\sigma_3 = \sigma_4 , \tag{7}$$

$$J_{13} = J_{24} , \tag{8}$$

$$J_{14} = J_{23} . \tag{9}$$

must be valid, as will be shown below.

The two *gauche* conformers of Fig. 8 are mirror images and hence must have equal populations ($P_b = P_c = P_g$). The population of the *trans* conformer ($P_a = P_t$) will approach equality with P_g in the high temperature limit only. The effective screening constants are given by:

$$\sigma_1 = P_t \cdot \sigma_1^a + P_g(\sigma_1^b + \sigma_1^c), \tag{10}$$

$$\sigma_2 = P_t \cdot \sigma_2^a + P_g(\sigma_2^b + \sigma_2^c). \tag{11}$$

Analogous expressions hold for the screening constants σ_3 and σ_4. By symmetry, $\sigma_1^a = \sigma_2^a$ and $\sigma_1^b = \sigma_2^c$ and hence Eq. (6) follows from Eqs. (10) and (11). By the same reasoning, σ_3 is shown to equal σ_4.

The average values of the spin couplings J_{ij} are similarly given by the expression:

$$J_{ij} = P_t \cdot J_{ij}^a + P_g(J_{ij}^b + J_{ij}^c). \tag{12}$$

Now, by symmetry, the following relationships must hold:

$$\begin{aligned} J_{13}^a &= J_{24}^a\,; \quad & J_{14}^a &= J_{23}^a \\ J_{13}^b &= J_{24}^c\,; \quad & J_{14}^b &= J_{23}^c \\ J_{13}^c &= J_{24}^b\,; \quad & J_{14}^c &= J_{23}^b\,. \end{aligned} \tag{13}$$

When Eqs. (13) are inserted into the Eqs. (12) corresponding to $i=1$ or 2 and $j=3$ or 4, one obtains immediately the equalities (8) and (9). This completes the proof of the symmetrical equivalence. The protons of a symmetrically equivalent pair will in general not be magnetically equivalent unless *all* the vicinal spin couplings are equal, and such equality cannot be inferred from symmetry considerations alone. If one assumes, however, as is often done, that the vicinal coupling in any one rotamer depends only on the dihedral H–C–C–H angle, then all the couplings of Eq. (13) may be represented by two different couplings, *viz.* J_g, the coupling between gauche positioned nuclei and J_t, the coupling between *trans* positions. In such case the motionally averaged values of the spin couplings will be given by

$$J_{14} = P_t \cdot J_t + 2P_g \cdot J_g = J_{23} \tag{14}$$

and

$$J_{13} = P_t \cdot J_g + P_g(J_t + J_g) = J_{24} \tag{15}$$

and magnetic equivalence will then be attained in the high-temperature limit when $P_t = P_g$.

II.2. Time Averaging Phenomena

a) General Effects of Time-Dependent Interactions

In previous sections it has been repeatedly mentioned that observed NMR parameters are to be interpreted as resulting from motional averaging. It is important to consider the time scales involved in these time-averaging phenomena.

From the theory of Fourier analysis it is known [5] that a phenomenon of duration τ is represented by a Fourier frequency distribution of a width not less than $\delta\nu \simeq 1/2\,\pi\tau$. The reader may have encountered an example of this relationship in the laboratory, where frequency measurements of a precision $\delta\nu$ require a measuring time of the order of $1/\delta\nu$. The Heisenberg uncertainty relation

$$\Delta \mathrm{E} \cdot \Delta t \geqq \hbar \tag{16}$$

is obviously equivalent to the inequality

$$\delta\nu \cdot \Delta t \geqq 1/2\,\pi\,. \tag{17}$$

In NMR one of the consequences of the conditions (16) and (or) (17) is that an interaction, in order to produce two resolved lines separated by a frequency $\Delta\nu$, must have a duration τ that is large compared to $1/2\,\pi\Delta\nu$. If the duration τ is comparable to $1/2\,\pi\Delta\nu$, the spectral features caused by the interaction will be smeared out over a width of the order of $\delta\nu = \Delta\nu$. Finally, if the duration of the interaction is short compared to $1/2\,\pi\Delta\nu$, only the average value of the interaction will affect the energy levels and transition frequencies of the spectrum, but the time-dependent modulation of the interaction will still produce line broadening effects that may be observed in the spectrum unless obscured by stronger extraneous line-broadening effects.

Example 1: Motional Averaging of Dipolar Interactions

Many examples of time-averaging phenomena are encountered in NMR spectroscopy. The direct "through space" magnetic interactions of nuclear dipoles in a solid produces structure of a width that may amount to some 10 to 20 KHz.

A line-width of 10 to 20 KHz is usually large enough to smear out all structure caused by chemical shifts and indirect nuclear spin-spin couplings (J couplings); consequently high-resolution NMR is almost impracticable unless the effects of dipole-dipole interactions are annihilated or grossly diminished, either by molecular tumbling in fluids or by some experimental expedient.

The isotropic average of the dipole-dipole interaction vanishes and, in liquids, where the correlation time for molecular reorientation is of the order of 10^{-11} sec, only a very weak line-broadening effect – normally obscured by extraneous effects, in particular by magnetic field inhomogeneity – remains of that interaction. If a nonspherical molecule is dissolved in a liquid crystal matrix, the motional averaging may become slightly anisotropic and in that case direct dipole-dipole interactions may produce additional structure in a high resolution NMR spectrum. The analysis of the NMR spectra from such partially oriented molecules will not be discussed in the present book.

Example 2: Motional Averaging of Chemical Shifts by Reorientation

The currents induced in an electron cloud of a molecule in general depend upon the orientation of the molecular framework with respect to the external H_0 field, and the observed screening constants of the nuclei are always motionally averaged values. The correlation time pertinent to this particular averaging process is the correlation time of molecular rotation and is typically of the order of 10^{-11} sec in a liquid. Often the times involved in other motional averaging processes are considerably larger, as exemplified next.

Example 3: Motional Averaging of Chemical Shifts by Ring Inversion

In cyclohexane, the rotational motion of the molecule should produce two different averaged screening constants, one for the axial and one for the equatorial protons. At room temperature, however, rapid chair-to-chair interconversion interchanges axial and equatorial protons at a rate estimated to be of the order of 10^5 sec^{-1}. The chair-chair interconversion can be slowed down sufficiently at low temperatures ($\sim -80°$ C) to allow separate determination of the equatorial and axial proton chemical shifts, and it is found that the chemical shift difference $\Delta\nu$ is of the order of 30 Hz at a spectrometer frequency of 60 MHz.

b) Limiting Cases of Very Slow and Very Rapid Rates

Theories have been developed that allow the calculation of the complete NMR absorption profile under the influence of time-dependent phenomena. The limiting cases of very long and very short time constants give simple expressions for the line-broadening effects of the interaction. If a particular NMR energy level has a life time of the order of τ, all the lines originating in that level will be broadened by an amount $\delta\nu = 1/2\pi\tau$ in the limit of long τ values. In the limit of *short* τ values, the coalesced lines of the time-averaged spectrum will have a line width contribution of the order of $2\pi(\Delta\nu)^2\tau$, where $\Delta\nu$ represents a measure (the "second moment") of the spectral width that would have been found in the absence of time averaging. These formula may be used for estimating the limitations on resolution caused by time-dependent interactions.

Example 4: Collapse of Spin Multiplets through Proton Exchange

The spectra of aliphatic alcohols are modified by hydroxyl proton exchange. The spin coupling between the hydroxyl proton and a vicinal C–H proton is of the order of 7 Hz. If the residence time of the hydroxyl proton in any particular molecule is greater than 0.2 sec ($2\pi J\tau > 10$), the spin-spin splittings will still be resolved in the spectrum but the multiplet lines are broadened. Unless particular precautions have been taken in preparing the sample, impurities that catalyze the proton exchange are normally present in sufficient amounts to reduce the residence time τ of the hydroxyl protons in a given molecule to values below 0.01 sec, in which case spin-spin splittings caused by spin coupling to the hydroxyl protons are no longer observable. The collapse of the spin multiplets occurs both in the OH and CH absorption bands of the molecule. The splittings vanish because the average of $J_{AB}m_Am_B$ as seen from nucleus A vanishes, since the exchanging nuclei B come in equally often with $m_B = +1/2$ and with $m_B = -1/2$. The average of $J_{AB}m_Am_B$ as seen from B vanishes for the same reason.

Example 5: Collapse of Spin Multiplets by Quadrupole Relaxation

Spin couplings between a proton (or some other spin 1/2 nucleus) and nuclei with spin $I > 1/2$ are often rendered unobservable by the rapid quadrupole relaxation of such nuclei.

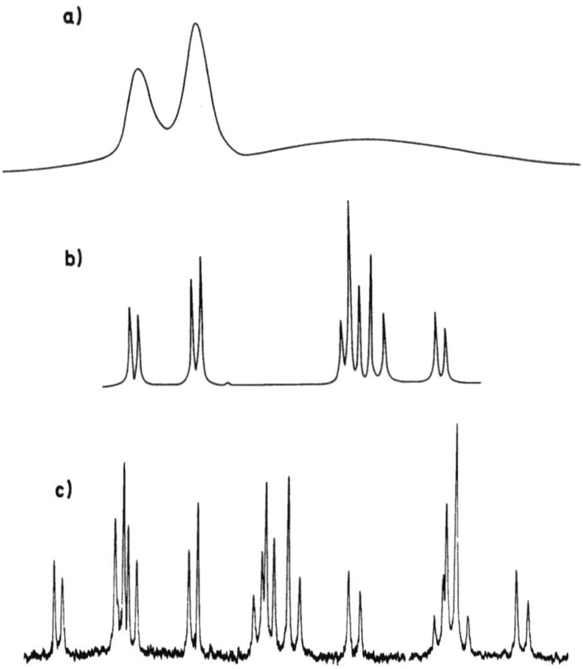

Fig. 9. Proton NMR spectra of formamide at 60 MHz. a Ordinary NMR spectrum of ^{14}N-formamide. The lines are broadened to coupling of the protons to the rapidly relaxing ^{14}N nucleus. b Double resonance spectrum of ^{14}N-formamide with strong irradiation of the ^{14}N resonance. c Ordinary NMR spectrum of ^{15}N-formamide

The interactions between the non-spherical charge distribution of nuclei with $I > 1/2$ and randomly varying inhomogeneous molecular electric fields produce a rather efficient mechanism for inducing thermal transitions between the magnetic spin states of such nuclei. When the rates of these thermally induced transitions become large compared to the spin-coupling constant, the spin-coupled proton only experiences the vanishing average value of this interaction over the $2I + 1$ spin states of the $I > 1/2$ nucleus (Fig. 2). At intermediate relaxation times of the $I > 1/2$ nucleus, a spin-coupled proton experiences a line broadening only. Such line broadening produced by the combined effects of spin coupling and quadrupole relaxation of a spin-coupled nucleus are common in proton resonance spectra of molecules containing nitrogen or boron[4]. An example of ^{14}N-quadrupole relaxation broadening in proton NMR is offered by the case of formamide (Fig. 9). The proton resonance spectrum of normal formamide displays only a broad absorption band containing little structure (Fig. 9a), whereas the molecule formamide-^{15}N produces a well resolved proton resonance spectrum, owing to the absence of quadrupole interactions of the ^{15}N nucleus ($I = 1/2$) (Fig. 9c).

[4] The quadrupole relaxation rate depends not only on the nuclear quadrupole moment but also on molecular structure and on the strength and rate of intermolecular interactions. Resolved H–^{14}N spin-coupling constants have been observed for example in several alkylisonitriles, in protonated alkylamines and in quarternary alkylammonium ions.

The quadrupole relaxation of the nuclei of the halogen atoms Br, Cl and I (but not of fluorine, which has spin 1/2) is found to be rapid enough to eliminate all evidence of proton halogen spin-coupling effects in proton resonance spectra of these halogenated compounds.

Example 6: Collapse of Spin Multiplets on Double Irradiation

Chemical exchange and quadrupole relaxation of a nucleus X produce a collapse of multiplet structure in a spin-coupled nucleus A when the succession of different spin states of the nucleus X proceeds at a rate that is fast compared to J_{AX}. One way of interpreting the phenomenon of spin decoupling through irradiation with a strong r.f. field is by analogy with these phenomena. Under the influence of a strong irradiation field, the irradiated nucleus X undergoes rapid oscillation between its spin states; when these oscillations occur at a high rate compared to J_{AX}, the multiplet structure in the resonance of a spin-coupled nucleus A disappears when driven by a weak observation r.f. field. The analogy must not be pushed too far, however, because chemical exchange and relaxation are stochastic processes occurring completely at random: there is no way of telling at what moment a particular spin will make a transition under the influence of one of these processes. On the other hand, the motion of the ensemble of spins X under the influence of the strong irradiation field is highly regular, since all spins in the sample experience the same radiofrequency amplitude at any one moment (for this reason we prefer the term "oscillations" to the term "transitions" to describe this latter motion). The temporal and spatial coherence of the irradiation field makes its effects at weak and intermediate perturbations, entirely different from those of the random processes of chemical exchange or relaxation. An interesting modification of spin decoupling has been devised by ERNST [6] who introduced random noise decoupling. With his device, which is now commercially available, the time coherence of the irradiation field is virtually eliminated and the effects of noise decoupling on the NMR spectrum show many similarities with the effects of chemical exchange. However, spatial coherence remains, since all the spins of the sample are surrounded by the same transmitter coil and some interesting differences remain.

II.3. The Energy Level and Transition Diagram

In Figs. 2, 4, and 5 of part I we have encountered diagrams representing the energy levels of a simple spin as a function of the H_0 field. In Fig. 10a the energy levels of two independent chemically shifted spins, A and X, are represented in the manner of Figs. 4 and 5 of Section I.2. Now, if the spins A and X belong to the same molecule, it is more appropriate to consider the *total* magnetic energy of the AX spin system. In the present case, when no spin coupling has to be taken into account, the total energy E of the spin system is simply the sum of the individual spin energies. It follows from Eqs. (9) and (10) of Section I.2 that the magnetic energy of this uncoupled two-spin system may be expressed in terms of the Larmor frequencies ν_A and ν_X and can be written as

$$E = h(\nu_A m_A + \nu_X m_X)\,. \tag{18}$$

Since the magnetic quantum numbers m_A and m_X can take up the values $+1/2$ or $-1/2$ independently of one another, we obtain four energy levels in all and, since the difference between ν_A and ν_X is small compared to their absolute values (chemical shifts are usually in the ppm range), the separation between the two energy levels that have m_A and m_X of opposite signs, becomes very small and has been grossly exaggerated in Fig. 10b. These two levels are in fact almost degenerate.

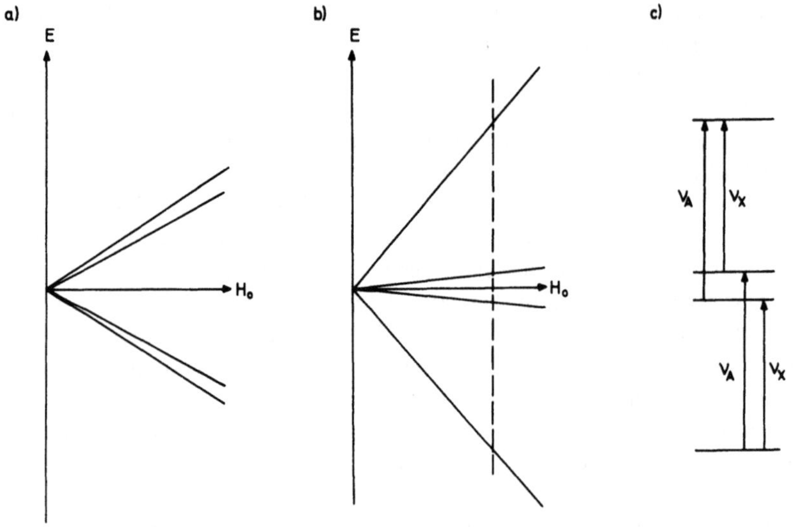

Fig. 10. a The energy levels of two independent chemically shifted nuclei as a function of the magnetic field H_0. b The *total* energy of an AX spin system as a function of the magnetic field H_0. c The energy levels and transitions at a fixed magnetic field of an AX spin system. The level displacements due to the relative chemical shift of the A and X nuclei are grossly exaggerated

In a frequency sweep NMR spectrum, the H_0 field is held constant and the energy of any one level is strictly constant. When discussing the structure of an NMR spectrum it is simpler to consider the spectrum as arising in this way. In ordinary single resonance spectra, the field sweep spectrum is indistinguishable from a frequency sweep spectrum. This is so because the total width of the sweep (H_0) over the spectrum is always very *small* compared to H_0 itself, and on a relative scale the variation in the energy levels caused by variation in H_0 over the spectrum is completely negligible.

Hence, in the following we shall consider all NMR absorption lines as arising from transitions between fixed energy levels. In Fig. 10c these fixed energy levels of the AX system depicted in Fig. 10b are represented by horizontal bars. Each bar corresponds to an intersection of the dotted vertical line through H_0' of Fig. 10b with the oblique energy curves; a vertical energy axis is implied in Fig. 10c, but no significance should be attached to the horizontal axis. The resonance of spin X at frequency ν_X is depicted in Fig. 10c as arising from two different transitions and the same is true for the A spin resonance. This is in agreement with our molecular point of view: every energy level corresponds to a stationary state of the spin system and we may conceive of our sample as being composed of four classes of molecules such that all the molecules of one class exist in the same stationary state. A transition in which m_X changes may involve *either* a molecule in which $m_A = +1/2$ *or* one in which $m_A = -1/2$ and, similarly, a transition of spin A may involve either a molecule with $m_X = +1/2$ or one with $m_X = -1/2$.

If the nuclei A and X are now spin-coupled with $|J_{AX}| \ll |\nu_A - \nu_X|$, the pairs of degenerate lines at ν_A and ν_X are split by an amount J_{AX} to produce a four-line AX spectrum. In the "weakly coupled" (cf. Section II.4) AX system, the spin coupling energy is represented by a term $E_j = h J_{AX} m_A m_X$ and the outermost levels, corresponding to equal signs of m_A and m_X of Fig. 10c, will obtain an energy increment of $hJ_{AX}/4$ while the two intermediate levels will have an energy contribution of opposite sign. The displacements of the energy levels of an AX spin system through spin-coupling effects are illustrated schematically in Fig. 11.

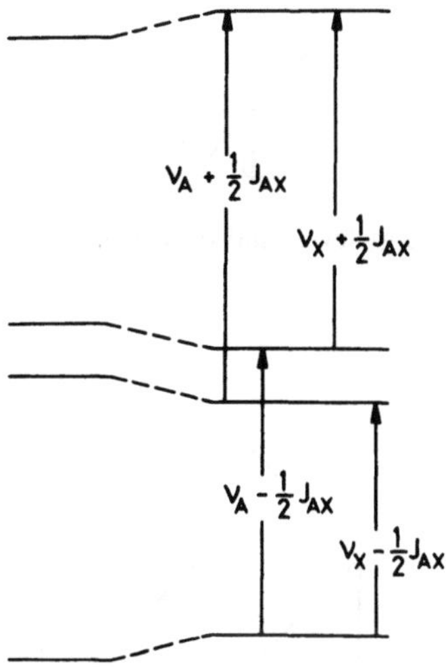

Fig. 11. The effect of a finite spin coupling J_{AX} on the energy levels and transitions in an AX spin system. The level shifts are grossly exaggerated

In this figure the energy displacements caused by chemical shifts and spin couplings are again grossly exaggerated to bring out the causes of line splittings. The assignment of the different transitions in Fig. 11 to changes in spin states can be made more perspicuous if the nearly degenerate energy levels are drawn next to each other rather than above each other. The resulting energy level and transition diagram is depicted in Fig. 12.

In Fig. 12 lines slanting downwards to the right correspond to transitions in which the spin state of X changes, and lines slanting downwards to the left represent A transitions. Equality of the splitting in the A and X parts of the spectrum follows from the frequency sum rule

$$\nu_{12} + \nu_{24} = \nu_{13} + \nu_{34} = \frac{E_1 - E_4}{h} \tag{19}$$

which may be rewritten

$$\nu_{24} - \nu_{13} = \nu_{34} - \nu_{12} \, . \tag{20}$$

In larger spin systems, where the energy-level and transition diagram contains many more levels and transitions, the horizontal bars used in Fig. 12 to indicate energy levels are sometimes left out since they represent an unnecessary complication of the figure. The graphical representation of the energy-level and transition

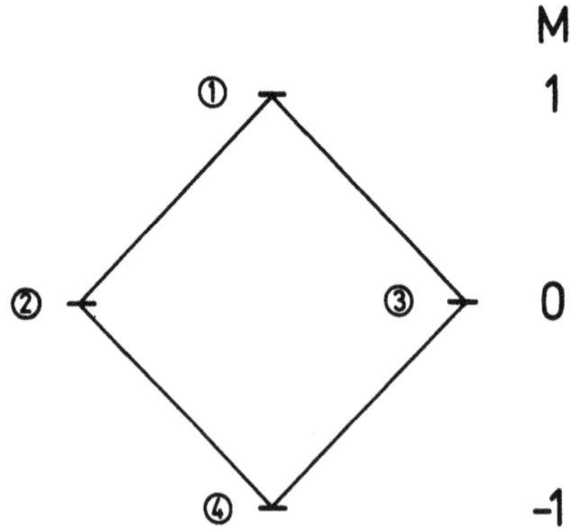

Fig. 12. The energy levels and transitions in an AX spin system

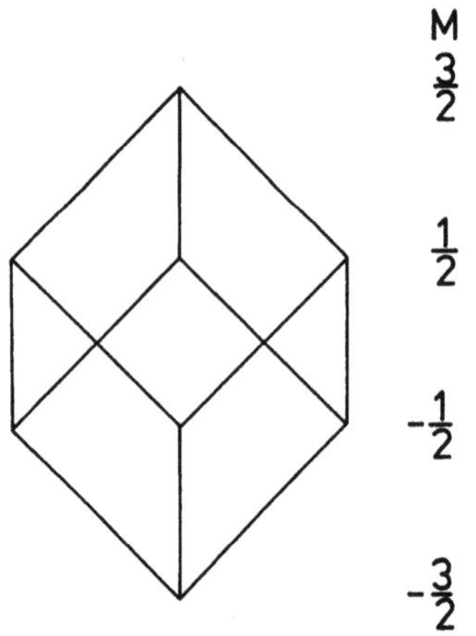

Fig. 13. The energy levels and transitions in an AMX spin system

diagram then takes the form of interwoven sets of polygons, each polygon corner representing an energy level and each edge representing a possible NMR transition. The energy level and transition diagram of a weakly coupled three-spin system AMX represented in this manner is depicted in Fig. 13.

Obviously, a large number of frequency sum rules can be found by simply summing the frequencies of any closed polygon, keeping in mind that the energy increases along the vertical axis of the figure.

When dealing with complex NMR spectra, it is often necessary to begin the analysis procedure by a trial and error search for repeated spacings to help in the assignment of observed lines to transitions in an energy-level diagram. The important property of the energy-level and transition diagram in this context is its topology and hence alternative representations may be found useful [7].

II.4. Strong and Weak Coupling. Nomenclature

a) Basic Ideas

A spin system of two protons A and X (spin 1/2) will have altogether four spin states. In an external magnetic field H_0 the lowest energy of this spin system will be obtained when the two protons are aligned with the H_0 field, *i.e.* when they both have magnetic quantum numbers $m_A = m_X = -1/2$, to yield a resulting magnetic quantum number $M = m_A + m_X = -1$; the highest energy corresponds to $M = +1$ and the two states with $M = 0$ both have an energy that is close to the one observed in the absence of the external field (Fig. 14). In the absence of spin coupling between the two nuclei, the energy separation between the two $M = 0$ levels is proportional, at any given H_0, to the chemical shift between the two protons and vanishes with vanishing chemical shift.

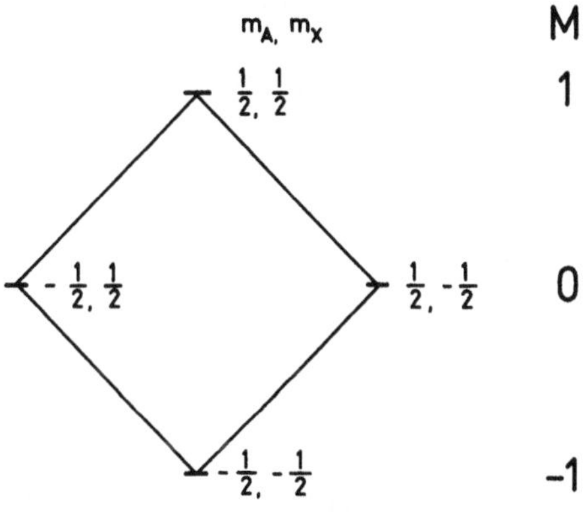

Fig. 14. The energy levels and transitions in an AX spin system. The numbers next to each level show the magnetic quantum numbers m_A and m_X in the different states, while the resulting magnetic quantum number M is given to the right of the diagram

In the presence of spin coupling, the spins are subject to a mutual interaction in addition to their interaction with the external H_0 field. Introduction of new interactions generally modifies the stationary states and energy levels of a system but, if this new interaction is weak, the modification is usually small. In order for a weak interaction to cause a significant change in the character of a stationary state of the original unperturbed system, the energy of this state must be nearly degenerate with that of another stationary state, so that the weak perturbing interaction may overcome, as it were, the tendency of the original strong interaction to preserve the individuality of its stationary states.

The resultant magnetic quantum number M is a characteristic of a stationary state *both* in the *original* system with uncoupled spins and in the *perturbed*, spin-coupled, system. However, the individual magnetic quantum number *m* of any one spin need not be preserved in a stationary state of the coupled-spin system. Thus, even though the spin-coupling energy always represents only a small interaction energy, as compared with interactions of any one spin with the external H_0 field, it may still modify (mix) the two states with $M=0$ which are nearly degenerate in energy. In the strong coupling limit, when the spin-coupling constant is large in comparison with the difference in Larmor frequency of the two coupled spins, the character of a (mixed) state with $M=0$ will resemble that of one original $M=0$ state no more and no less than that of the other $M=0$ state. The observed NMR transitions in the strong coupling limit cannot be viewed as flips of the spin of one particular proton, but must be viewed as a transition between two stationary states of the entire spin system. This statement may be formulated in terms of *selection rules:* in the *weakly coupled* two-spin system, the selection rule requires that the magnetic quantum number *m* of *one (and only one)* spin must change by unity in every observable transition. When spin-coupling effects are strong enough to modify (mix) the stationary states of the spin system, this simple rule breaks down. The *general selection* rule in NMR spectroscopy is that the *total* magnetic quantum number M of the coupled spins must change by unity:

$$\Delta M = \pm 1 \tag{21}$$

in a transition induced by a weak r.f. field.

Spin coupling among nuclei represents a tendency of the coupled nuclei to behave in their magnetic interactions with external fields as a single entity: only insofar as other interactions distinguish between the spin-coupled nuclei can their individuality be observed in an NMR experiment. The terms *strong* or *weak coupling* are used to indicate which one of these two tendencies prevails in a given spin system. Thus magnetically equivalent nuclei are always strongly coupled to each other but may be either strongly or weakly coupled to other groups of nuclei.

Interactions that tend to distinguish between different spin-coupled nuclei are, for example, differences in their interaction with electronic currents, manifested in unequal screening constants and Larmor frequencies or differences in their spin couplings to a third, chemically shifted, nucleus. Since differences in Larmor frequency increase in proportion to the applied H_0 field, the structure of a high-resolution NMR spectrum often becomes simpler when the spectrum is obtained at a higher value of H_0, *i.e.* at a higher value of the nominal spectrometer frequency.

b) Weak Coupling in a Simple Spin System

When the Larmor frequency ν_i of nucleus *i* is compared with the Larmor frequency ν_j of a spin-coupled nucleus *j*, it may be the case that the difference $|\nu_i - \nu_j|$ is large compared to the magnitude of the spin-coupling constant J_{ij}. Many authors have chosen to speak of "weak coupling" between the nuclei *i* and *j* when this situation prevails. The condition is, however, neither a necessary nor a sufficient one for the spectrum to become simpler ("first order") in appearance.

We prefer to present first a *sufficient* condition for a simple system in the present subsection and to formulate a general (and somewhat abstract) definition of weak coupling in subsection (c) below. The general definition will be clarified with some useful criteria for establishing the occurrence of weak coupling in subsection (f) below after a brief discussion of magnetic quantum numbers in subsection (d) and a presentation of the notations for spin systems in subsection (e).

If *every* spin coupling constant J_{kl} in a many-spin system is small in magnitude compared to *every* chemical shift difference ($v_i < v_j$), then the spins of that system are all weakly coupled to one another and produce a simple first-order spectrum (cf. Section IV.1 below). Every resolved line in the NMR spectrum of such a spin system may properly be assigned to the transition of only one spin.

c) Weak Coupling in a General Case. The Concepts of Basic Groups and Basic Multiplets

In an N-spin system the resultant *z* component F_z of the angular momentum of the entire spin system is obtained by adding the contributions of all the spins (A, B ... N) according to

$$F_z = I_z(A) + I_z(B) + \dots I_z(N) \tag{22}$$

F_z is always a "constant of the motion" and assumed a definite value, the resultant magnetic quantum number M of the system, in *every* energy level of the spin system, irrespective of the extent of coupling among the spins.

In addition the magnetic quantum numbers $m_A, m_B \dots m_N$ of the individual spins are also all defined in *every* stationary state, provided that *all* the spins of the multi-spin system are *weakly* coupled to one another.

There exist many spin systems in which *some* spins; A, B and C, say, are strongly coupled among themselves but weakly coupled to the *rest* of the spins. The magnetic quantum numbers m_A, m_B and m_C of the individual spins in such a group *g* are *not* well-defined in every stationary state, but the resultant *z* component $F_z(g)$ of angular momentum *of the group g*, defined by

$$F_z(g) = I_z(A) + I_z(B) + I_z(C) \tag{23}$$

assumes a definite value (the magnetic quantum number M_g *of the group g*) in every stationary state of the spin system. To a given energy level it must be possible to assign one and only one value of M_g. The group *g* is then said to constitute a *basic group* of the spin system.

The decomposition of a large spin system into basic groups thus follows from the sets of magnetic quantum numbers that are definable in every stationary state of the system. We have chosen to make the term of "weak coupling" depend on the same criterion: the spins of a basic group are always weakly coupled to spins outside the group, but more or less strongly coupled among themselves. (If the spins of a group were found to be weakly coupled among themselves the decomposition should have been carried further in order to yield the *basic* groups.)

For example: if M, the resultant magnetic quantum number of the entire spin system, is the *only* magnetic quantum number that is well defined in every energy level of the system, then the entire system forms one single basic group and all spins of the system are more or less strongly coupled to one another. On the other hand, if the magnetic quantum number m_i of every spin i is well defined in every stationary state, then every spin forms a basic group by itself.

When a spin system may be decomposed into basic groups, the NMR spectrum of that system can always be decomposed into *basic multiplets* and the NMR absorption band of any basic multiplet arises from transitions in which the magnetic quantum number M_g of the corresponding basic group changes by unity. That is to say, the general *selection rule* of Eq. (21) is supplemented by the more stringent condition that the resultant *magnetic quantum number of one and only one basic group must change by unity* in every observable transition.

In large organic molecules it often happens that some nuclei, though spin-coupled among themselves, are not measurably coupled to nuclei in other parts of the molecule. The spectrum then decomposes into completely independent partial spectra. This trivial decomposition is an extreme case of decomposition into basic multiplets.

The concepts of strong and weak coupling are limiting concepts only, and in practice a group g is considered a basic group if some particular value of $F_z(g)$ for all practical purposes determines the character of every particular state. The question as to where the borderline should be drawn is obviously related to the desired accuracy of the NMR parameters of the system investigated. In part, that question will be answered by Chapter IV.

d) The Magnetic Quantum Numbers of Groups of Spins

The magnetic quantum number m of a single spin I can assume one of the $2I+1$ values obtained by going from $-I$ to $+I$ in integer steps, as mentioned in Section I.1. In a strongly coupled spin system, the magnetic quantum numbers m_i of the individual spins need not assume definite values in a stationary state, but the resultant magnetic quantum number M is a characteristic of any stationary state; the same is true for M_g, the magnetic quantum number of a basic group g (in short: M and M_g are "*good quantum numbers*"). Since these resultant magnetic quantum numbers retain their values independently of the extent of coupling within the group, their possible values are most readily arrived at by going to the limit of vanishing coupling where they are obtained by adding the individual contributions from the constituent spins.

Thus in the two-spin system considered in Fig. 12, the resultant magnetic quantum number M was found to have one of the values $+1$, 0, and -1, and the statistical weight of the $M=0$ state was twice that of the $M=+1$ or $M=-1$ states. More generally, the resultant magnetic quantum number of a group of n nuclei with spin 1/2 can assume all values from $-n/2$ to $+n/2$ in integer steps, and the number (or the statistical weight) of states with $M=-n/2+p$ is given by the number of ways in which p nuclei with $m=+1/2$ and n-p nuclei

Table 2. *The statistical weights (number of states) for a total magnetic quantum number M in a system of nuclei of spin $I=1/2$. The rows of the table are arranged according to n and the columns according to M. A blank entry implies that the corresponding M-value cannot be attained with the chosen value of n. The filled entries form Pascal's triangle, in which every number is obtained as the sum of its two nearest neighbours in the row immediately above*

	−5	−9/2	−4	−7/2	−3	−5/2	−2	−3/2	−1	−1/2	0	1/2	1	3/2	2	5/2	3	7/2	4	9/2	5
0											1										
1										1		1									
2									1		2		1								
3								1		3		3		1							
4							1		4		6		4		1						
5						1		5		10		10		5		1					
6					1		6		15		20		15		6		1				
7				1		7		21		35		35		21		7		1			
8			1		8		28		56		70		56		28		8		1		
9		1		9		36		84		126		126		84		36		9		1	
10	1		10		45		120		210		252		210		120		45		10		1

with $m = -1/2$ may be picked out from the batch of n nuclei. This number is given by the binominal coefficient $\binom{n}{p}$ and is listed in Table 2. Note that odd values of n produce only half-integer values of M, and even values of n produce only integer values of M.

e) Notations for Spin Systems

The line structure in a high-resolution NMR spectrum depends only on the relative magnitudes and signs of the NMR parameters – no matter what structural features of the molecule may be held responsible for their numerical values. The basic analysis procedure depends primarily on the manner in which the spins may be arranged into basic groups and sets of equivalent nuclei. The notation system to be described below was originally developed as an aid in this classification by POPLE, SCHNEIDER and BERNSTEIN and is designed for spin systems containing only spin 1/2 nuclei. In the early NMR literature the distinction between different types of equivalence was not always made. Slight modifications of the conventions presented below may be found in the literature because different authors have introduced different auxiliary symbols (asterisks, primes etc.) to indicate the *type* of equivalence among equivalent spins.

Every nucleus of the spin system is represented by a capital letter A, B ... to Z. Nuclei within one basic group are represented by consecutive (or identical – cf. below!) letters, usually chosen from the beginning of the alphabet, for the group in which one is primarily interested. Nuclei that are weakly spin-coupled to this group are represented by letters further away in the alphabet: P, Q, ... X, Y, Z.

Thus, ABC represents a spin system with only three nuclei that are strongly coupled among themselves, ABX represents a spin system with only three nuclei, two of which, A and B, form a strongly coupled pair that is weakly coupled to nucleus X, and ABXY represents a four-spin system of two basic groups AB and XY with strong coupling among the nuclei within each group and weak coupling between the groups. In the ABPX four-spin system, spins P and X are each weakly spin-coupled to the AB pair and also weakly coupled among themselves. In sum: the basic groups of ABX are AB and X, those of ABXY are AB and XY and those of ABPX are AB, P and X.

A group of magnetically equivalent nuclei is designated by a single letter with a lower index integer representing the number of spins in the group. For example, A_3B_2X represents a six-spin system with two strongly coupled groups of magnetically equivalent nuclei that are weakly coupled to the single spin X. The two basic groups of A_3B_2X are: A_3B_2 and X.

The different spins of a symmetrically equivalent group are all represented by the same capital letter but the individual spins are distinguished by means of primes. Thus AA′XX′ represents a four-spin system in which A and A′ are symmetrically equivalent, as are X and X′, and the mutual coupling between the A nuclei and the X nuclei is weak (two basic groups). AA′BB′ represents a similar four-spin system in which the coupling between the A and B nuclei is strong. The AA′BB′ spin system constitutes one single basic group.

No particular notation is required for designating chemical shift equivalent nuclei because chemical shift equivalence by itself does not simplify the spectrum of a general spin system. Some authors, however, have confounded the notation system by using primes and identical letters for different chemical shift equivalent nuclei as for symmetrically equivalent nuclei. However, we shall use for example the notation ABX to designate a three-spin system of two basic groups AB and X, even when A and B are two chemical shift equivalent nuclei that are not magnetically equivalent.

f) Rules for Establishing the Occurrence of Weak Coupling

It now remains to present criteria for distinguishing between weak and strong coupling. First, it may be said that in a heteronuclear spin system containing magnetic nuclei of different isotopic species, all the nuclei of one isotopic species will be weakly coupled to nuclei of other isotopic species under normal operating conditions, $H_0 > 1000$ gauss, say.

Furthermore, when comparing the relative magnitudes of Larmor frequency differences and spin couplings (cf. subsection (b) above), it is usually sufficient to consider the individual ratios $|J_{ij}/(\nu_i - \nu_j)|$. Only if the spins under consideration, i and j, are coupled to a third nucleus k in such a way that $J_{ik} - J_{jk}$ is comparable in magnitude with $|\nu_i - \nu_j|$ can strong coupling between nuclei i and j occur, even though the ratio $|J_{ij}/(\nu_i - \nu_j)|$ is small.

When a spectrum displays an isolated absorption band of a width that is small compared to its separation from neighbouring absorption bands, and if the integrated intensity of the isolated band is found to correspond to an integral number of nuclei, one may conclude that the isolated band is in all probability[5] a basic multiplet and arises from transitions ($\Delta M_g = 1$) of a basic group in the spin system. The converse is by no means true: a basic multiplet may, for example, overlap with absorption bands assigned to transitions of other spins that are not spin-coupled to the nuclei of the basic group g.

II.5. Invariants in NMR Spectra

In Section I.4 we have already encountered one important invariant of NMR spectra, *viz.* the integrated intensity of a basic multiplet which is known to be proportional to the number of nuclei in the basic group to which the transitions in the multiplet are assigned. The integrated intensity of the basic multiplet thus is independent of the extent of coupling within the spins of the basic group. Other invariants may be found that are helpful in the analysis of complex spectra, even when strong coupling effects make the spectra deviate from first order.

a) The Spectral Moments

One important invariant is the chemical shift at the *center of gravity* of a basic multiplet, which is completely independent of *all* spin coupling effects and

---

[5] Exceptional cases where this rule does break down may be found in Section IV.5.e, Example 1.

depends only on the arithmetic mean of the chemical shifts of the spins belonging to the corresponding basic group. If the basic multiplet possesses mirror-image symmetry, the arithmetic mean of the chemical shifts of the spins of the basic group is thus obtained on inspection. If mirror-image symmetry is lacking, it is a computational advantage if the spectral data output is obtained in digital form. It must be emphasized, however, that this simple center of gravity rule holds only for a basic multiplet and is valid only to the extent that the group assigned to the multiplet is weakly coupled to other groups.

Other invariants that are sometimes found useful are the higher spectral moments of a basic multiplet. The q : *th* moment of a spectrum with respect to an arbitrary frequency origin is defined as

$$\langle v^q \rangle = \frac{\sum_{mn} (v_{mn})^q \cdot S_{mn}}{\sum_{mn} S_{mn}} . \tag{24}$$

Where v_{mn} is the NMR absorption *frequency* corresponding to the transition between the stationary states m and n (referred to that origin) and S_{mn} is the signal strength produced by that transition as measured by the area under the peak. With $q = 1$, Eq. (24) gives the center of gravity of the multiplet.

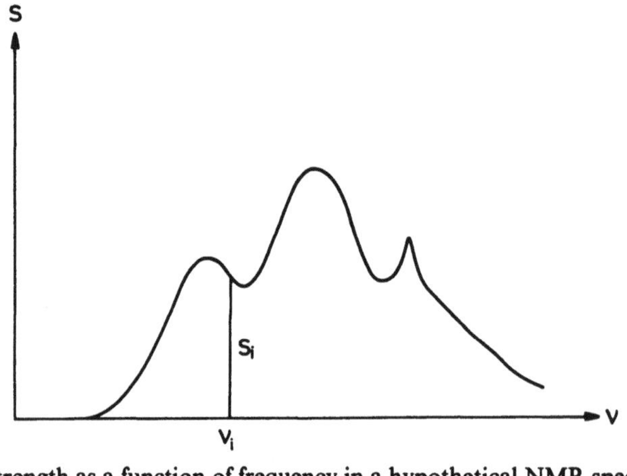

Fig. 15. Signal strength as a function of frequency in a hypothetical NMR spectrum. S_i and v_i are used in evaluating the spectral moments

In practical cases, the relative intensities of the individual transitions may be obscured by overlap of resonance lines and, in that case, formula (24) is replaced by a similar formula in which v_{mn} is exchanged for v_i, the frequency at the i : th of an arbitrarily chosen set of closely speced frequency intervals, and S_{mn} is exchanged for S_i, the signal amplitude at frequency v_i (Fig. 15). Again, digital output of spectrometer data will expedite the processing according to Eq. (24).

It will be noted that the definition (24) makes the spectral moment independent of units for line intensities.

The dependence of the higher spectral moments on the NMR parameters is simplest and most practically useful in spin systems where all magnetic nuclei form one single basic group. When referred to the center of gravity as a frequency origin, the second and third moments of the entire spectrum depend only on the differences in Larmor frequency of the nuclei. Since all the three lowest moments in this case depend only on the chemical shift, it follows that the chemical shifts of spin systems described by no more than three different screening constants may in principle be obtained from moment calculations, *regardless of the values of spin coupling constants.*

Example 1: Screening Constants of Three Chemical Groups

Consider, for example, an $AA'BB'C_3$ spin system. Since the first three spectral moments do not depend on the spin couplings, their predicted values are most readily obtained by arbitrarily setting all spin coupling constants equal to zero. The spectrum would then consist of three peaks with relative intensities 2:2:3 centered at the Larmor frequencies ν_A, ν_B and ν_C (assumed unknown). The center of gravity $\bar{\nu}$ would be given by

$$\langle \nu^1 \rangle = \bar{\nu} = \frac{2\nu_A + 2\nu_B + 3\nu_C}{7} \tag{25}$$

and the second and third moments with respect to the center of gravity ("central moments") would be given by

$$\langle (\nu - \bar{\nu})^2 \rangle = \frac{2(\nu_A - \bar{\nu})^2 + 2(\nu_B - \bar{\nu})^2 + 3(\nu_C - \bar{\nu})^2}{7} \tag{26}$$

and

$$\langle (\nu - \bar{\nu})^3 \rangle = \frac{2(\nu_A - \bar{\nu})^3 + 2(\nu_B - \bar{\nu})^3 + 3(\nu_C - \bar{\nu})^3}{7}. \tag{27}$$

In Eqs. (25) to (27) the left-hand members are experimentally accessible when defined in terms of the observed quantities of the right-hand side of Eq. (24). Thus three equations are obtained for the three unknowns, ν_A, ν_B, and ν_C.

In a spectrum containing two or more basic multiplets already, the second moment of a basic multiplet spectrum becomes dependent on spin-coupling constants and, in fact, explicit expressions for the higher spectral moments of basic multiplets in terms of the NMR parameters have only been developed for spin systems containing two basic groups.

By means of the powerful tools of transformation theory and group theory, many invariants of spin systems and NMR spectra have been derived as aids in spectral analysis. Most of these invariants are, however, not as readily recognized in actual NMR spectra as is, for example, the center of gravity of a basic multiplet. For this latter reason, a comprehensive survey of all known invariants is not felt to be in the spirit of the present text and the interested reader is referred to the specialized literature (cf. Section VI).

However, the mode of approach to the analysis problem throughout parts IV and V leans heavily upon the concepts of such invariants as: the symmetry types of spin states and transitions in spectra of symmetrically equivalent nuclei, the "composite particle" concept in the analysis of spectra of magnetically equivalent nuclei, and the division of a complicated NMR spectrum into sub-spectra.

III. Miscellaneous Aids for Spectral Assignment

III.1. The Need for Additional Information

A common method for analysing NMR spectra proceeds as follows. For a given type of spin system, theoretical formulae are derived for the frequencies and relative intensities of all the allowed transitions. On the basis of chemical intuition, supported by correlation with related compounds, a set of trial NMR parameters is chosen and a trial NMR spectrum is calculated by use of the theoretical formulae. By comparing the observed and calculated spectra one may obtain an assignment of the observed transitions.

If this assignment is the correct one and if the trial parameters were reasonably close to those of the compound observed, an iterative procedure may usually be applied for adjusting the trial parameters until the best fit is achieved.

For the simple spin systems to be discussed in the following chapter, the technique just described will not be followed. Instead the analysis proceeds by the identification of certain types of subspectra to be discussed in connection with the various spin systems below. By simple algebraic operations on the observed frequencies within such subspectra, the values of the NMR parameters may be derived. All these calculations may be performed by means of a slide rule or, for greater precision, with a desk-top calculator. The main difficulty in this approach is to identify the subspectra in the often complex and far from "perfect" NMR spectra encountered in practice.

Ambiguities may arise for several reasons. First, it may happen that a number of transitions, that in principle are allowed, have such a low transition probability as to be lost in the background noise under normal operating procedures. Secondly, the number of actually observed lines may be lower than that theoretically expected because of near or perfect coincidence of two or more theoretically distinct transitions. Finally, one feature often employed for the identification of subspectra is the occurrence of repeated spacings (frequency intervals between pairs of lines), and hence the accidental occurrence of repeated spacings may represent an additional source of ambiguity.

Obviously, improvements in spectrometer performance reduce the number of ambiguities in most cases. Weak signals are less apt to escape detection with improved signal-to-noise ratio. Accidental overlap of neighboring transitions becomes less probable with improved spectrometer resolution and nearly equal frequency intervals may be distinguished with improved precision in the frequency measurements. The information content may, however, be increased not only by improving the quality of the spectrum but also by supplementing a given spectrum with spectra obtained under other operating conditions. The most important modifications in operating conditions from our present point of view are:

(i) operation at a higher level of the stimulating radio frequency field H_1;

(ii) a change in NMR parameters, primarily in the Larmor frequencies;

(iii) the use of double irradiation techniques.

III.2. The Effects of a Strong Stimulating Field

The amplitude H_1 of the stimulating r.f. field employed in recording NMR spectra is often set low enough to ensure the absence of all saturation effects in the spectrum. Under such operating conditions, the relative intensities of two transitions are strictly proportional to the corresponding transition probabilities. With increasing power level, the lines that have the largest transition probabilities are the ones most readily saturated. The maximum achievable intensity of a line is proportional to the square root of its transition probability, and the lines that are strong in the unsaturated spectrum attain this maximum at a lower power level than do the weak lines. It must be emphasized that this rule holds only for lines that are weak owing to a low transition probability and does not apply to lines that are weak owing to low abundance. Thus impurity bonds and lines derived from less common isotopic species, such as ^{13}C satellites in proton resonance spectra, saturate just as readily as their counterparts in more abundant species. Spinning and modulation sidebands, on the other hand, are less readily saturated than the fundamental spectrum.

From what has just been said, it follows that weak transitions may often be made observable by running a new spectrum at a higher r.f. field. The method has several limitations though, because lines that have a very low transition probability will never become strong enough to be observable, and the intensities of weak lines that do become observable through the use of a strong r.f. field do not usually bear a simple relationship to the corresponding transition probability owing to onset of saturation. Furthermore, one must guard against the possibility that spurious lines caused by spinning side-bands may be mistaken for genuine transitions in the spectrum. To distinguish between spinning side-bands and genuine transitions, it is advisable to study the spectrum at different spinning rates.

A second phenomenon associated with strong r.f. fields is the occurrence of *double quantum* (or two quanta transitions). In contrast to the normal (or *single quantum*) transitions, the double quantum transitions obeys the selection rule

$$\Delta M = \pm 2 \quad (1)$$

and the frequency of such a transition obeys a modified Bohr frequency rule

$$\Delta E = 2h\nu . \quad (2)$$

Thus, a double quantum transition will always occur midway between two single quantum transitions[6]. The relationship between the single quantum and double quantum transitions may be visualized as in Fig. 16. The energy difference of the two terminal levels having total magnetic quantum numbers M and M − 2 is spanned by three pairs of single quantum transitions *ab*, *cd*, and *ef*, all having a common center frequency, and the double quantum transition, which is indicated by a dashed line in Fig. 16, will appear at this centre, as follows from Eq. (2). The

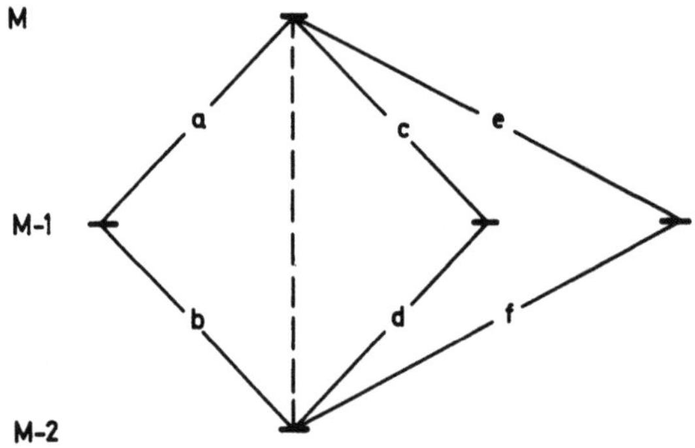

Fig. 16. Energy-level diagram showing three pairs of progressively connected transitions. The double quantum transition is indicated by the broken line

two single quantum transitions within any one pair, for example, *a* and *b*, are *connected* by a common energy level and the connection is said to be *progressive* because the energy of the common level is intermediate between those of the two terminal transitions. Similarly, two transitions, such as *a* and *c* or *b* and *d*, are *connected* by a common energy level but this connection is said to be *regressive* because the common energy level is either higher or lower than any of the two other levels involved in that pair of transitions. Thus, a double quantum transition must always occur midway between progressively connected transitions. Conversely, the occurrence of a double quantum transition may be used to establish the existence of pairs of progressively connected transitions. These may be quickly located in the normal single quantum spectrum by placing one leg of a pair of dividers at the position of the double quantum line and finding which single quantum lines have a partner symmetrically placed on the other side of the double quantum transition.

Usually double quantum transitions cannot be observed until the r.f. power is sufficient to saturate the single quantum lines, and the double quantum transition thus appears as a sharp peak on a broader background. Well below its own saturation level, a double quantum transition increases in intensity with the third power of the r.f. amplitude, in contrast to a single quantum line that

---

[6] Owing to higher order effects the double quantum transitions may sometimes be shifted away frow this midpoint. In practice this shift is usually quite small and of the order of a tenth of a Hertz [*8*].

increases linearly with H_1 below saturation[7]. A double quantum transition will have a large transition probability (and hence appear at a comparatively low value of r.f. power), provided that the two lines in one pair of progressive transitions are close together in frequency and yet do not overlap. The double quantum peak then appears midway between these two single quantum lines, unless they differ very much in intensity, in which case the double quantum line tends to be displaced towards the stronger of the two single quantum lines. This shift grows with the square of the r.f. amplitude but usually does not attain values in excess of a few tenths of a Hertz before the double quantum line has become saturated.

To sum up: spectra recorded at increased r.f. amplitudes may be employed (*i*) to locate weak transitions in the normal spectrum, and (*ii*) to identify pairs of progressively connected transitions with the aid of double quantum transitions.

III.3. Controlled Variations of the Larmor Frequencies

For a given nucleus in a given sample the Larmor frequency, as defined in Section I.2a, is strictly *proportional* to the applied magnetic field. Consequently, one may change the appearance of spectra from strongly coupled spin systems simply by recording the spectra at two or more different nominal spectrometer frequencies. The attraction of this method resides in the fact that the relationships between the NMR parameters in the different spectra are predictable and strictly under experimental control. A disadvantage of this method is that it requires expensive equipment. A poor man's substitute for the change-of-spectrometer method is the utilization of specific medium effects. As long as solvent-solute interactions do not lead to conformational changes, the specific medium interactions usually alter the spin couplings far less than the Larmor frequencies and, if the composition of the solution is changed gradually, there will normally be no difficulty in keeping track of the assignments. Solvents that have been found to produce particularly pronounced medium effects usually exhibit large magnetic anisotropies and/or tendencies to form association complexes, for example, through hydrogen bonding [*9*]. In particular, benzene or benzene-d_6 and pyridine are frequently used in this connection.

III.4. Double Resonance Methods

a) Basic Concepts of Double Resonance

Double resonance implies that the spin system study is subjected to two r.f. fields, known as the irradiation and the observation fields, respectively. The observation field H_1 will usually be weak enough to avoid saturation and the observation channel of the spectrometer will be made quite narrow and centered

[7] Certain types of NMR spectrometers incorporate an automatic gain control (AGC) which reduces the gain of the receiver in proportion to H_1 and this AGC must be accounted for or disconnected when one desires to compare line intensities at different irradiation levels.

on the frequency of H_1. The irradiation field will always have an amplitude H_2 strong enough to perturb appreciably one or more of the irradiated transitions. An irradiation field of amplitude H_2 may be considered to have an effective width of $\gamma H_2/2\pi$ Hz around its center frequency ν_2, so that lines having a transition frequency that differs from ν_2 by a few multiples of $\gamma H_2/2\pi$ will remain virtually unaffected by the irradiation field.

It is customary in double resonance work to distinguish between *homonuclear* and *heteronuclear* double resonance experiments, according to whether the irradiated and observed nuclei are of the same isotopic species or not. The experimental facilities required for performing a heteronuclear double resonance experiment are much more elaborate than those required for homonuclear double resonance where, in principle, an audiofrequency oscillator and a phase sensitive detector are all that one requires.

The method of scanning the spectrum is of fundamental importance in double resonance experiments, in contrast to what applies to single resonance experiments. In the *frequency sweep* mode of operation the magnetic field H_0 is kept constant, as is the radio frequency ν_2, so that the irradiation field is applied at a fixed spectral position and the spectrum is scanned by varying the frequency ν_1 of the observing r.f. field. In the so-called INDOR (*in*ternuclear *do*uble *r*esonance) mode of double resonance, the magnetic field H_0 is likewise constant but the observation field H_1 has its frequency centered at a single transition. The intensity of the selected signal is monitored while the perturbing field of variable frequency ν_2 traverses other regions of the spectrum. In the *field sweep* mode of operation, the frequency difference $\nu_1 - \nu_2$ is kept constant and the spectrum is scanned by varying the magnetic field H_0. The latter method implies that different parts of the spectrum are observed under different conditions of perturbation and yields results that are difficult to interpret unless the frequency interval $\nu_1 - \nu_2$ matches only one or at most a few frequency intervals in the observed single resonance spectrum.

Double resonance spectra may facilitate the analysis problem in two entirely different ways. In the first case the irradiation field is used to smplify the appearance of one basic multiplet by obliterating the spin couplings of the corresponding basic group with spins in a second basic group subjected to the strong spin decoupling irradiation. In the second case the irradiation field is applied selectively, preferably to a single non-degenerate transition, and serves as a probe for finding pairs of transitions that are *connected* through a common energy level. In this context it is again important to distinguish between *progressive* and repressive pairs of transitions (cf. Subsection III.2 above and Fig. 16).

b) Spin Decoupling

An ideal spin decoupling experiment requires that the widths of the two basic groups involved are both small compared with the difference in their center frequencies. When this condition is fulfilled, the frequency ν_2 of the irradiation field may be centered within the irradiated basic multiplet, and the effective width $\gamma H_2/2\pi$ of the irradiation field may be made large in comparison to the width of the *irradiated* basic multiplet and yet small enough to have a negligible interaction with the transitions in the *observed* basic multiplet. When this condition is fulfilled, the effects of the spin couplings between the irradiated basic group and other basic groups become obliterated, and the observed basic multiplets appear as if the irradiated basic group were absent. With somewhat weaker irradiation fields one may attain partial decoupling, for example, collapse of spin multiplet lines in certain subspectra, but in general a non-ideal spin decoupling will result in the occurrence of additional lines, sometimes called satellites, and deviations from the intensity relations of an ideal experiment. The detailed analysis of such spectra obtained at moderately strong irradiation amplitudes is somewhat complex, and often the information obtainable through such experiments may be obtained more easily by the selective irradiation methods to be discussed next.

c) Double Resonance Methods Based on Population Transfer

The intensity of an unsaturated NMR absorption line is proportional to the product of the probability for the transition and the population difference of the two stationary states involved in the transition. Under normal operating conditions the population difference is governed by the Boltzmann distribution and – because of the smallness of the radio frequency quanta – the population difference is the same for all the transitions of a given nuclear species.

If a transition is saturated by the application of a strong r.f. field, the populations of the two levels involved in the transition become equal. Since this situation is a non-equilibrium one, the thermally induced transitions (relaxation processes) work to restore the equilibrium, and this normally will lead to *secondary* population changes in levels other than those directly affected by the r.f. saturation. If the different thermal transition probabilities are all of the same order of magnitude, the largest population changes will, however, remain in the two levels that are directly involved in the transition saturated. On the basis of this insight, double resonance methods have been devised for identifying pairs of connected transitions and the kind of connection.

The clearest representation of the results is obtained by performing the experiment in the INDOR mode. The frequency ν_1 for the observation field is centered on a non-degenerate transition, *a*, say, and the amplitude H_1 is set low enough to cause little or no saturation of that line. The amplitude H_2 of the irradiation field is set strong enough to cause appreciable saturation of an irradiated line but low enough to render $\gamma H_2/2\pi$ less in magnitude than the resolving power, $\Delta\nu$, of the spectrometer so as to avoid spin "tickling" (cf. below). The spectrum is scanned by allowing the frequency ν_2 of the irradiation field to sweep *slowly* through the spectrum. The effects of this INDOR experiment will be explained with reference to the schematic energy level and transition diagram of Fig. 17a, the single resonance spectrum of Fig. 17b and the INDOR spectrum of Fig. 17c. The figures may be interpreted as fragments of the entire spectrum. The lower horizontal line in the INDOR spectrum of Fig. 17c represents the base line and the horizontal parts of the upper trace in Fig. 17c represent the undisturbed intensity of line *a* as obtained with the observation field H_1. As ν_2 passes through ν_1, a beat pattern arises because the observation channel will then

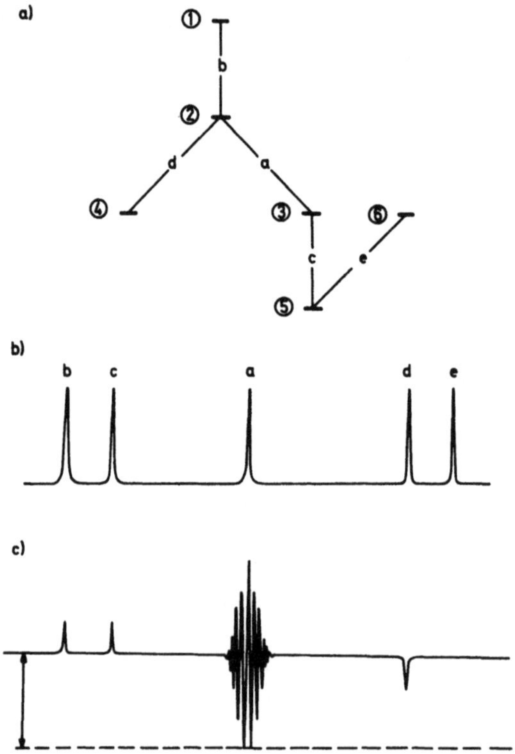

Fig. 17. a, b Schematic spectrum with five lines and the corresponding energy-level diagram. c The resulting INDOR spectrum, when the observation field H_1 is centered on transition *a* while the irradiation field H_2 is swept through the spectrum. The broken line shows the level of the baseline

accept the H_2 irradiation field. This beat pattern will disappear when ν_2 has receded away from ν_1 by more than the bandwidth of the detector. When ν_2 coincides with the resonance frequency of line *b*, the transition *b* becomes saturated and hence the population of the lower level, 2, involved in resonance *b* becomes depleted, because at equilibrium the population of level 2 exceeds that of level 1, and saturation results in an equalization of the populations of levels 1 and 2. Now, a depletion of level 2 implies that the excess population of level 3 over level 2 increases and hence the intensity of line *a* becomes enhanced. Analogous arguments lead to the conclusion that saturation of line *c* enhances the intensity of line *a*, though now this enhancement is brought about by an increase in the total population of the lower level 3 involved in the transition. It will be noted that both lines *b* and *c* are *progressively connected* with line *a*, and it follows that a saturation of a progressively connected line enhances the intensity of the monitored transitions. Saturation of the regressively connected transition *d*, on the other hand, reduces the intensity of line *a*, because the equalization of populations among levels 2 and 4 results in an increases in the population of level 2 and hence in a decrease in the excess population of level 3 over level 2. Finally, saturation of a line, such as *e*, that is not connected with the observed transition will, at most, result in an indirect population transfer to levels 2 and 3 and tends to become unobservably small even under favourable conditions.

Information similar to that obtainable from the INDOR experiment just discussed may be obtained in a less conspicuous way by means of a frequency sweep experiment with ν_2 centered at *a*. As ν_1 scans through the spectrum, one will first obtain a beat pattern when ν_1 passes through ν_2, thereafter lines *b* and *c* will be recorded with slightly enhanced intensities, line *d* will appear with a reduced intensity and line *e* will remain unperturbed, as in the single resonance spectrum. The information on the intensity changes in this mode of operation obviously can be obtained only by comparison of the observed single resonance and double resonance intensities.

Of all the double resonance methods conceivable, those based on population transfer are the most selective, because the irradiation amplitude required to produce observable effects are the lowest with such methods, hence the effective width of the perturbation becomes smallest.

d) Spin Tickling

When the irradiation amplitude H_2 is increased so as to render $\gamma H_2/2\pi$ comparable to the smallest splitting $\Delta\nu$ resolvable with the spectrometer, a new phenomenon, colloquially known as *spin tickling*, becomes important. Double resonance behaviour under spin tickling conditions may be stated in the form of three rules:

1. If a non-degenerate transition is selectively irradiated, all connected transitions will be split into doublets.

2. Of the observed splittings, those of regressive transitions are split into well resolved doublets and those of progressive transitions are split into broadened doublets.

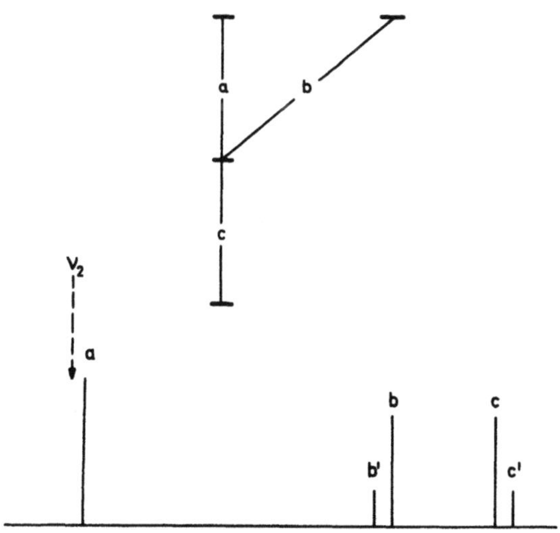

Fig. 18. Schematic diagram showing the splittings obtained in the *b* and *c* lines when the connected *a* transition is perturbed by an irradiation field H_2 of frequency ν_2

3. The doublet splitting is proportional to the square root of the intensity of the irradiated line and to the amplitude H_2 of the irradiation field.

In addition to these three rules, which are valid for the symmetrical doublets observed when ν_2 is centered exactly on a connected transition, the following rules pertaining to the case of unsymmetric tickling may be found helpful:

4. If a weak irradiation field is applied selectively in the neighbourhood of one single non-degenerate transition, but slightly off resonance, all connected transitions will be split into asymmetrical doublets.

5. The stronger line of such a doublet will nearly coincide with the position of the corresponding single resonance line, and the weak satellite will be displaced from that position in the same sense as the offset for a regressive transition but in the opposite sense for a progressive transition (Fig. 18).

The occurrence of a doublet splitting is usually better recognized in a symmetric tickling experiment but the distinction between progressive and regressive transitions, which may sometimes be ambiguous, is often better recognized by repeating the experiment with a slight offset, provided that the selectivity condition can still be maintained in the asymmetric case. This will become difficult if the line to be irradiated is not several line widths removed from its nearest neighbours, and in such cases one may prefer to use a population transfer experiment.

IV. Instant Analysis of NMR Spectra

IV.1. Simple First Order Spectra

Simple first order spectra arise from spin systems in which each basic group consists of one single spin or one single group of magnetically equivalent nuclei. The line structure of a basic multiplet can then be described by simple superposition of its spin-spin splittings.

Thus, the NMR absorption band of a basic group A_a of a magnetically equivalent nuclei in a first order spectrum is split by spin coupling to a second basic group K_k into q_k equidistant bands, or submultiplets, separated in frequency by J_{AK}, where q_k stands for the number of possible values of the magnetic quantum number M_k. (If the K nuclei all have spin 1/2, the number q_k equals $k+1$, cf. Table 2.) The relative intensities of these submultiplet bands are given by the statistical weights of the corresponding M_k values and are those listed in Table 2, if the K nuclei have spin $I = 1/2$. If the basic group A is spin coupled to a third group P_p, every basic submultiplet of the A – K interaction is further split into q_p equidistant bands separated in frequency by J_{AP} and with relative intensities corresponding to the statistical weights of the different M_p values, and so on.

The above rules may be rationalized by reference to the energy-level structure in a $A_aK_kP_p$ type spin system. The transition frequencies in the A basic multiplet, *i.e.* the frequencies of the transitions in which M_A changes by unity, are given by

$$\nu = \nu_A + J_{AK}M_K + J_{AP}M_P \,. \tag{1}$$

Eq. (1) may be derived in the same way as Eqs. (26) and (27) of Section I.3.

A particular pair of values for the magnetic quantum numbers M_K and M_P corresponds to one particular line in the A multiplet, and the intensity of this line is proportional to the number of molecules in the sample that have this set of values for M_K and M_P. The Boltzmann factors of the NMR energy levels are all nearly equal to unity owing to the weakness of magnetic interactions ($\Delta E \ll kT$), and hence the populations are almost entirely determined by the statistical weights.

It is readily shown that the appearance of a first order spectrum is completely independent of the absolute and relative signs of the spin couplings.

Example 1. 2-furanaldehyde

The 60 MHz proton resonance spectrum of 2-furanaldehyde is shown in Fig. 19. This spectrum is not strictly first order but very nearly so. The deviations from first order are most prominent in the intensity destribution in the multiplets assigned to H3 and H4, but are small enough to make a first order analysis valid to a very good approximation (cf. below). Every basic group in 2-furanaldehyde consists of only one nucleus and the spin coupling patterns in the basic multiplets

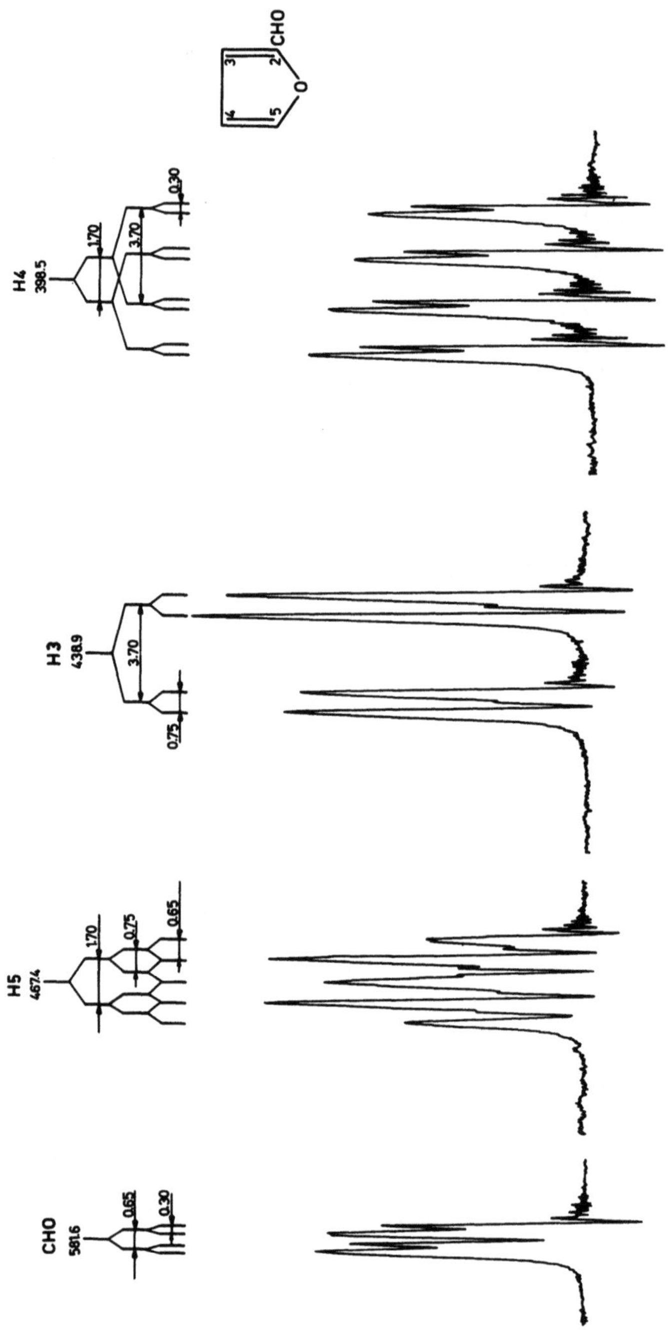

Fig. 19. NMR spectrum at 60 MHz of 2-furanaldehyde in 20% (by weight) CCl_4 solution. The chemical shifts are given relative to TMS as internal reference. The aldehyde band is shown with reduced (one-half) gain

are explained by simple branching diagrams above each multiplet. Thus the number 0.65 in the branching diagram of the CHO multiplet represents the magnitude (in Hz) of the spin coupling J_{CHO-5} and this number recurs in the basic multiplet of H5. Similarly, the spin couplings $J_{CHO-4} = 0.30$ Hz, $J_{45} = 1.70$ Hz, $J_{35} = 0.75$ Hz and $J_{34} = 3.70$ Hz are identified by the pairwise occurrent splittings. Since the basic multiplet of H4 is completely resolved into eight lines, each of these lines represents a non-degenerate transition and should be of unit intensity. Slight deviations from this intensity distribution are apparent in the observed spectrum and depend in part on transient effects (wiggles), which tend to depress the second line in every pair of the closely spaced (0.30 Hz) doublets, and in part on second order effects which tend to increase the intensity of the H4 lines closest to the H3 band at the expense of the intensities of the H4 lines further away from the H3 band. Similar second order effects caused by the coupling J_{34} are also observed in the H3 band, where the right hand doublet is stronger than the left hand doublet. Since J_{CHO-3} is vanishingly small, the intensity of any one H3 band should be approximately twice that of any H4 band, but this relationship is not very well borne out because of transient perturbation of the H4 intensities. The intensity distribution in the H5 multiplet is not appreciably multilated by transients nor influenced by second order effects and arises through overlap of individual lines; this overlap amounts to almost perfect degeneracy in the second and fourth multiplet line, but is less perfect in the centre line which appears broadened. The CHO band, which was recorded with reduced (1/2) gain, displays no second order effects but transients again distort its spectrum.

a) Second Order Effects in Near First Order Spectra

Perfect first order spectra are seldom obtained from homonuclear spin systems, and the question naturally arises what are the errors in the NMR parameters incurred through a first order analysis of a non-perfect first order spectrum? The ratios

$$\lambda_{ij} = \frac{J_{ij}}{(\nu_i - \nu_j)} \tag{2}$$

represent dimensionless quantities (perturbation parameters), the magnitudes of which determine the deviation from first order behaviour in near first order spectra. Thus, if the accuracy of the observed transition frequencies is no better than $\delta\nu$ and if the ratio $\delta\nu/J_{ij}$ is large in comparison with the ratio λ_{ij} for every i and j, then the observed transition frequencies will not deviate measurably from those predicted on a first order basis and, conversely, the NMR parameters derived by means of a first order analysis cannot be significantly improved by taking second order or higher order terms in account.

Second order corrections become simplest in spectra of spin systems in which every basic group contains only one nucleus of spin 1/2, as in the 2-furanaldehyde spectrum above. In that case, no additional lines occur through second order effects, and the splittings observed in the basic multiplets remain equal to the corresponding coupling constants to within a correction factor quadratic in the perturbation parameters (λ_{ij}^2 or $\lambda_{ij}\lambda_{kl}$). Thus, the line spacings *within* a basic multiplet remain unchanged at their first order values as long as the quadratic correction terms can be neglected, but the spacings *between* basic multiplets depend linearly on the perturbation parameters λ_{ij}. The spin coupling between two of the coupled nuclei, A and K, say, produces a shift of the whole A multiplet pattern by $J_{AK}^2/2(\nu_A - \nu_B)$ and of the K multiplet pattern by $J_{AK}^2/2(\nu_K - \nu_A)$. It will be noted that these second order shifts tend to increase the separation between the A and K multiplets beyond the value $|\nu_A - \nu_K|$. If the A and K nuclei are in turn spin coupled to other chemically shifted nuclei, the second order displacements of multiplet patterns caused by the different spin coupling are simply added together. Deviations from first order behaviour are most conspicuous in the intensity distribution of the multiplet lines. Thus, in a strict first order spectrum the multiplet pattern of any basic multiplet displays perfect mirror-image symmetry with respect to its center frequency; on the other hand, when two basic groups are sufficiently strongly coupled to make the perturbation parameter λ_{ij} non-negligible, this mirror image symmetry is lost in a manner which increases the intensities of those parts of the two bands which are nearest to each other at the expense of the intensities of the "outer" parts of the multiplets. This "roofing" effect is discernable in the H3 and H4 bands of 2-furanaldehyde in Fig. 19 and is of great diagnostic value in the analysis of NMR spectra when certain resonances are hidden under broad absorption bands. For example, if one basic multiplet with a slanting intensity distribution has been recognized in the spectrum, the direction and magnitude of the slanting may be used for an approximate estimate of where its spin-coupled partner is to be found in the spectrum.

In a spin system containing several magnetically equivalent nuclei in a basic group, the second order effects will result in further line splittings and the line spacings within a basic multiplet will then no longer depend linearly on the spin couplings. Examples of such spin systems will be discussed shortly.

The foregoing discussion of second order effects is not exhaustive and is meant primarily as an aid in estimating the precision with which the NMR parameters can be obtained by first order analysis when, in fact, slight deviations from first order structure are discernable in the spectrum. If, in a practical analysis situation, this precision is deemed insufficient it is recommended that a complete analysis be attempted. Explicit expressions for the transition frequencies to third order and of relative intensities to terms linear in the perturbation parameter λ_{ij} are available in standard texts and many simple spin systems amenable to exact analysis are dealt with in the subsequent sections of the present chapter.

Example 2: Second Order Effects in the 2-furanaldehyde Spectrum

In the spectrum of 2-furanaldehyde depicted in Fig. 19, the largest perturbation parameter is found to be $\lambda_{34} = 3.70/40.4 = 0.092$. This value is large enough to make second order effects visible in the spectrum, where the high-field doublet of H3 should have its intensity increased by 9.2% from its first order value at the expense of the intensity of the low-field doublet. The predicted intensity ratio $1.09/0.91 \simeq 1.20$ agrees well with the experimentally observed value. The second order shift of the H3 and H4 multiplets is only $(3.70)^2/80.8$ Hz $= 0.17$ Hz and is almost negligible. The observed splittings coincide with the true coupling constants to within $(0.092)^2 \simeq 0.8\%$.

Example 3: The Proton Resonance Spectrum of 5-methyl-4-thiolen-2-one(I)

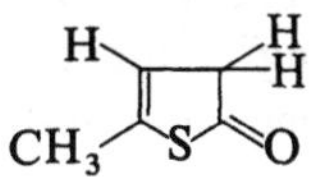

(I)

The 60 MHz proton magnetic resonance spectrum of (I) displays no second order effects, and even the 40 MHz spectrum reproduced in Fig. 20 has the appearance of a simple first order spectrum. The methyl group is split into three doublets with intensities 1 : 2 : 1, with a doublet splitting of 1.5 Hz caused by coupling to the ethylenic proton and a triplet spacing between doublets of 2.5 Hz representing the coupling between the methyl and $-CH_2-$ protons.

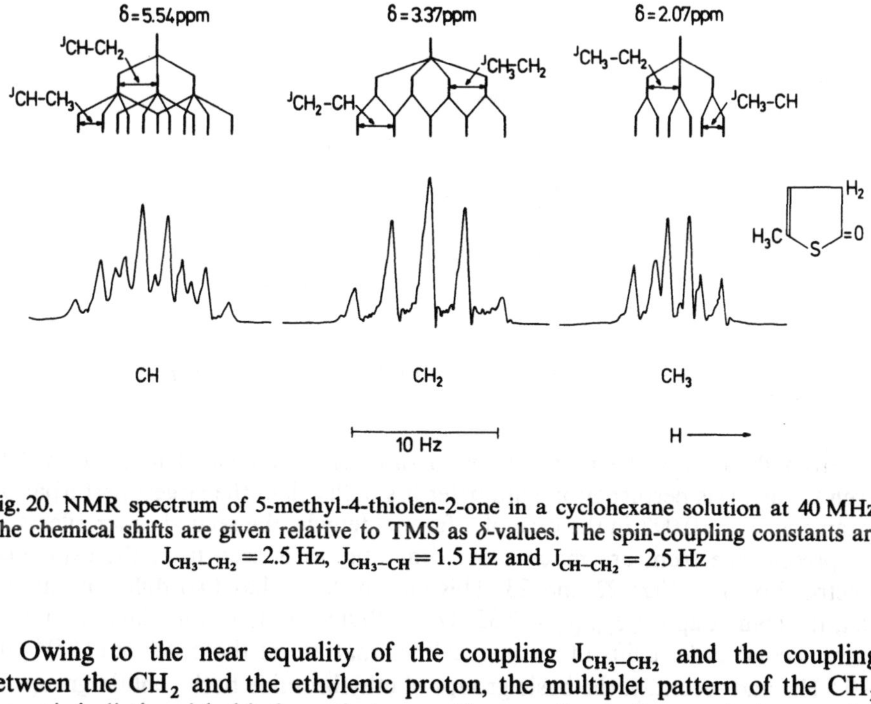

Fig. 20. NMR spectrum of 5-methyl-4-thiolen-2-one in a cyclohexane solution at 40 MHz. The chemical shifts are given relative to TMS as δ-values. The spin-coupling constants are $J_{CH_3-CH_2} = 2.5$ Hz, $J_{CH_3-CH} = 1.5$ Hz and $J_{CH-CH_2} = 2.5$ Hz

Owing to the near equality of the coupling $J_{CH_3-CH_2}$ and the coupling between the CH_2 and the ethylenic proton, the multiplet pattern of the CH_2 group is indistinguishable from that caused by coupling to four equivalent nuclei. Thus it becomes a quintet with relative intensities close to 1 : 4 : 6 : 4 : 1.

The ethylenic hydrogen resonance consists of a triplet of quartets. The triplet character arises from spin coupling to the CH_2 group, and the intensities of the three quartets are in the ratio 1 : 2 : 1. The quartet structure arises from the spin coupling with the methyl group and the lines within each quartet have an intensity ratio of 1 : 3 : 3 : 1. The two innermost quartet lines of the two outer quartets of the ethylenic proton band are not resolved in Fig. 20.

The largest perturbation parameter in this spin system has a magnitude of $2.5/52 \simeq 0.048$ and causes second order shifts of the order of 0.12 Hz.

Example 4: Ethylbromide

The 60 MHz spectrum of neat ethylbromide, CH_3CH_2Br, is shown in Fig. 21 and displays some "roofing" indicative of second order effects, but the 1 : 3 : 3 : 1 structure of the methylene band and the 1 : 2 : 1 structure of the methyl band are still easily recognized.

On a straightforward first order basis, the chemical shift obtained from this spectrum is for the CH_3 group $\delta = 1.67$ ppm and for the CH_2 group $\delta = 3.46$ ppm. From splittings as measured in a large number of spectra, like that in Fig. 21, the spin coupling is estimated on a first order basis to have a value of 7.35 Hz. The reproducibility of this average value is ± 0.1 Hz.

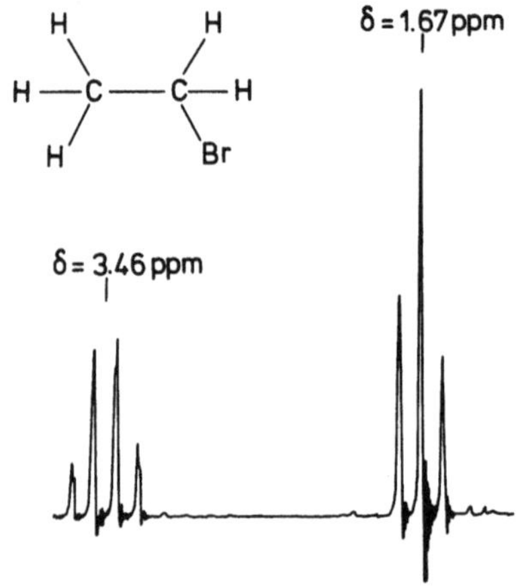

Fig. 21. NMR spectrum at 60 MHz of neat ethylbromide

From these data the perturbation parameter is calculated to be $\lambda = 0.068$. Even though this perturbation parameter is smaller than the largest perturbation parameter $\lambda_{34} = 0.092$ in the case of 2-furanaldehyde, the second order effects in the present spectrum are much more dramatic, as is seen from the expanded spectra shown in Figs. 22 and 23. This circumstance has two different causes: first, the spin coupling $J_{CH_2CH_3} = 7.35$ Hz in ethylbromide is quite large and produces second order shifts of resonance lines that are of the order of $(7.35)^2/108 \simeq 0.5$ Hz, and secondly, the basic groups of ethylbromide both contain groups of magnetically equivalent nuclei, so that additional splittings of the order of 0.5 Hz will arise in the multiplet lines through second order effects. Nevertheless, the frequency intervals indicated in Figs. 22 and 23 coincide with 3J and 2J, respectively, to within a correction term of the order of λ^2, and the error in the values so obtained 7.35 ± 0.05 Hz due to neglect of higher order effects is only $\lambda^2 \simeq 0.5\%$. Thus the "first order" estimate of the spin-coupling constant in the present case is correct to within the experimental error.

The accepted value of $J_{CH_2CH_3}$ in ethylbromide, as obtained through the exact A_2B_3 analysis of the spectrum, is 7.33 Hz. Our first order estimates of the chemical shifts should have an analytical error of the order of 0.5 Hz, but the

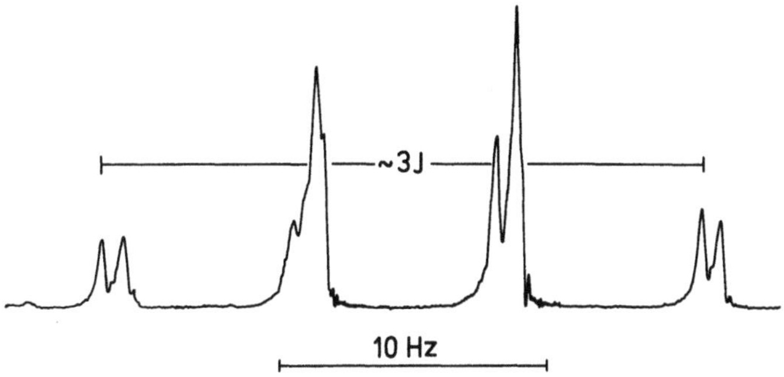

Fig. 22. Expanded spectrum of the CH_2 band in ethylbromide

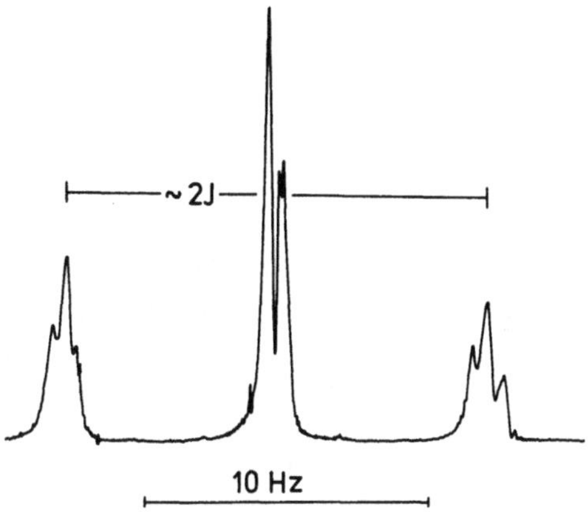

Fig. 23. Expanded spectrum of the CH_3 band in ethylbromide

difference from the published values valid for 10% solutions in CCl_4 is more than this, probably due to medium effects, which normally affect chemical shifts more than spin couplings.

b) Accuracies of First Order Estimates in A_nX_m Spin Systems

The procedure exemplified in the analysis of the ethylbromide spectrum can be generalized to any A_nX_m spectrum. The first order estimates of chemical shifts will always be in error by an amount of $J_{AX}^2/|(\nu_A - \nu_X)|$, whereas spin couplings may be estimated with a relative accuracy of λ_{AX}^2. The recommended procedure is best discussed separately for the AX_m, A_2X_m and A_3X_m cases, respectively.

The two outermost multiplet lines of the first order A multiplet spectrum of an AX_m system remain unsplit by second order interactions, and their separation corresponds to mJ_{AX} to a relative accuracy of λ_{AX}^2.

The two outermost multiplet lines of the first order A_2 multiplet spectrum of an A_2X_m system split into 1 : 1 doublets under the influence of second order interactions (cf. the CH_2 multiplet of ethylbromide in Fig. 22), and the separation between their centers corresponds to mJ_{AX} to a relative accuracy of λ_{AX}^2.

In an A_3X_m spectrum the two outermost multiplet lines of the first order A_3 multiplet spectrum split into 1 : 2 : 1 triplets under the influence of second order interactions (cf. the CH_3 multiplet of ethylbromide in Fig. 23), and here the separation between their centers again equals mJ_{AX} to a relative accuracy of λ_{AX}^2.

The above rules may be found helpful, since they offer a quick method for obtaining good estimates of the NMR parameters in not too strongly coupled A_nX_m spectra. They are not strictly required, however, because even strongly coupled A_nB_m spectra can always be analysed exactly by the aid of published tables, a desk calculator and a moderate amount of patience (cf. Section IV.7 below).

IV.2. The Two-Spin System AB

a) The General Structure of an AB Spectrum

An AB spin system is defined as a system of two-spin coupled nuclei A and B, both of spin $I = 1/2$, and with Larmor frequencies ν_A and ν_B such that $\lambda_{AB} = J_{AB}/(\nu_A - \nu_B)$ is not negligibly small. The AX spin system is one limiting case of the AB system ($\lambda_{AB} \to 0$), A_2 representing the opposite limit ($\lambda_{AB} \to \infty$). The AB analysis procedure to be described here is applicable over the entire range.

The theory of the AX spectrum was developed by retaining only the *z* component of the spin vectors, but from the discussion in Section II.4 it should be apparent that those two states of an AB spin system which correspond to $M = 0$ become modified under the influence of the spin coupling interaction in a way that precludes identification of a particular set of individual magnetic quantum numbers, m_A, m_B with a stationary state. In other words, the *z* components $I_z(A)$ and $I_z(B)$ no longer adequately represent the actual spin vectors $\boldsymbol{I}(A)$ and $\boldsymbol{I}(B)$. When the transverse components of the spin vectors have to be considered, the energy levels and the spectrum of the AB system cannot be calculated without a more detailed quantum mechanical treatment (to be performed in part V), but the results of the calculation can be summarized in a few simple formulae (Eqs. (3) to (6) below).

Before listing these formulae, we will first discuss the structure of an AB spectrum in relation to the energy level and transition diagram. Every AB spectrum is a symmetrical four-line pattern displying a "roofing" effect, as exemplified by the spectrum depicted in Fig. 24. The transition and energy level diagram corresponding to the same AB spectrum is depicted in the left part of Fig. 24. By simple application of the frequency sum rule (cf. Section II.3), it follows that the highest and lowest frequency transitions in the observed spectrum must have a common energy level with $M = 0$, and that the two innermost lines originate in transitions involving the other $M = 0$ level. We have arbitrarily assigned the numbers (2) and (3) to these levels for future reference. It also follows from the frequency sum rule that the spacing between the two high-frequency lines must equal that of the two low-frequency lines.

It has earlier been said in Section II.4 that a particular transition in a strongly coupled spin system must not be viewed as arising from a change of the magnetic state of one particular nucleus, but we have nevertheless labelled the four lines in Fig. 24 A1, A2 and B1, B2, respectively, as if it were possible to

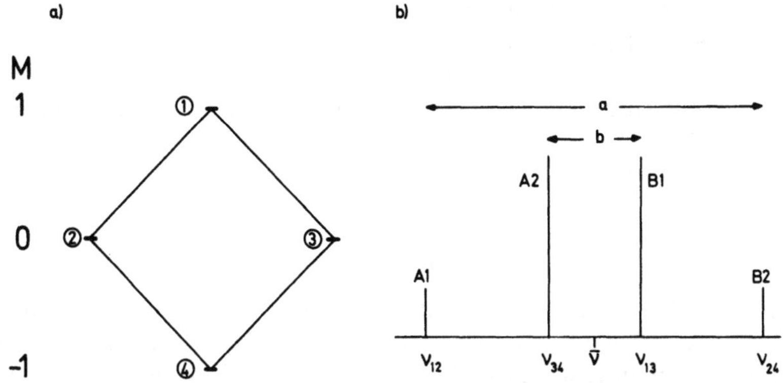

Fig. 24. Energy-level diagram and schematic spectrum of an AB spin system

assign a line either to an A or to a B transition. This is in accordance with common practice, where labels appropriate to a weakly coupled spin system are retained also in the general case. In the AB case, this assignment can be made simply by following the individual lines in a hypothetical series of experiments in which H_0 and consequently the difference $|\nu_A - \nu_B|$ gradually increase beyond the limits. In more complex spin systems, it is also possible to assign "first order" labels to all the observed transitions, but this assignment cannot always be done in a unique way and, furthermore, some of the transitions so assigned do not occur in the first order spectrum. These remarks should underline the conventional character of the assignments made. We hasten to add, however, that this arbitrariness applies *only* to the *naming* of transitions and that the NMR parameters obtained through a rigorous analysis always correspond to actual molecular properties.

b) Expressions for the NMR Parameters

The NMR parameters in the AB system are simply obtained from the frequency intervals a and b and the center frequency $\bar{\nu}$ indicated in Fig. 24. The appropriate expressions are given by

$$|\Delta| = |\nu_A - \nu_B| = \sqrt{ab}\,, \tag{3}$$

$$|J_{AB}| = \frac{a-b}{2}\,, \tag{4}$$

$$(\nu_A + \nu_B)/2 = \bar{\nu}\,. \tag{5}$$

A remarkable feature of the AB spectrum is that the coupling constant J_{AB} is always equal to the line splittings in the A1, A2 and B1, B2 pairs, irrespective of the chemical shift differences. Only the magnitude of J_{AB} can be obtained from the spectrum, but this is not a unique feature of the AB spectrum because, even in multispin systems, a simultaneous sign reversal of all the spin couplings produces no observable features in the spectrum. The sign of $\nu_A - \nu_B$ also remains undetermined, but this ambiguity is trivial since it depends only on the labels assigned by the spectroscopist.

As a check of consistency, one may wish to determine the relative intensities in the spectrum. The rule is simple: the intensities of the inner lines relative to those of the outer lines are in the ratio a/b. Expressed as an equation, this becomes with $I_{13} = I_{34} = I_a$ and $I_{12} = I_{24} = I_b$:

$$\frac{I_a}{I_b} = \frac{a}{b}. \tag{6}$$

As mentioned above, the two weak outermost lines originate in a common energy level (2) with $M=0$ and the two stronger lines have the common energy level (3). The difference in intensity is to be attributed to the difference in character of the two mixed $M=0$ states and does not depend on any inequality in the level populations. In the state (3) the spins A and B are coupled in a fashion that makes them behave in consonance in their interactions with the stimulating r.f. field, but in (2) the interaction of one spin with the r.f. field tends to cancel that of the other. At the limit, when $(\nu_A - \nu_B) \to 0$, the spins are completely coupled to form a resultant angular momentum, characterized by a total spin quantum number $F=1$ for the state (3) and $F=0$ for the state (2). The state of vanishing total angular momentum ($F=0$) has no magnetic moment and therefore cannot interact with a r.f. field. At this limit, therefore, the intensities of the two outermost lines vanish.

c) Limiting Cases of AX and A_2

It may be worth while to consider the appearance of the spectrum at the two limits, when $b/a \to 1$ and when $b/a \to 0$, which correspond to the two limits $|J_{AB}|/(\nu_A - \nu_B) \to 0$ and $|J_{AB}|/(\nu_A - \nu_B) \to \infty$ respectively.

At the limit when $b/a \to 1$, one obtains $a - b \ll a$ and consequently $2|J_{AB}| \ll a$ according to Eq. (4). When Eq. (4) is inserted into Eq. (3) and use is made of the inequality $2|J_{AB}| \ll a$, one obtains

$$|\nu_A - \nu_B| = \sqrt{a(a-2|J_{AB}|)} = a\sqrt{1 - \frac{2|J_{AB}|}{a}} \simeq a - |J_{AB}| \tag{7}$$

where the last member follows from a Taylor series expansion

$$\sqrt{1+x} = 1 + x/2 - x^2/8 + \dots \tag{8}$$

in which only the first two terms have been retained, since $x = 2|J_{AB}|/a \ll 1$. The first and last members of Eq. (7) may be combined with Eq. (4) to give

$$|\nu_A - \nu_B| = \frac{a+b}{2} \tag{9}$$

which is the expression used to obtain the chemical shift in a first order spectrum. If the third term in the series of Eq. (8) had been retained, one would have obtained the second order correction terms to the AX chemical shift expression.

The opposite limit, when $b/a \to 0$, corresponds to a situation in which $(\nu_A - \nu_B)/J_{AB} \to 0$, as may be shown in the following way. Since b is negligible in comparison with a Eq. (4) reduces to

$$|J_{AB}| \simeq a/2 \tag{10}$$

and, when (10) is inserted into Eq. (3), one obtains for b

$$b = \frac{(\nu_A - \nu_B)^2}{a} \simeq \frac{(\nu_A - \nu_B)^2}{2|J_{AB}|}. \tag{11}$$

Combining Eqs. (10) and (11) one obtains

$$b/a = \frac{(\nu_A - \nu_B)^2}{4(J_{AB})^2}. \tag{12}$$

Thus the limit $b/a \to 0$ implies that the ratio $(\nu_A - \nu_B)/|J_{AB}| \to 0$.

At the limit $(\nu_A - \nu_B) \to 0$, the AB spectrum degenerates into one single line at $\bar{\nu}$, *i.e.* into an A_2 spectrum, at a rate *much faster* than that in which $(\nu_A - \nu_B)$ vanishes, because the splitting of the two center lines b vanishes with the *square* of $\nu_A - \nu_B$ (cf. Eq. (11)), and the intensities of the outer lines, as compared to the intensities of the inner lines, vanish with the *square* of the ratio $(\nu_A - \nu_B)/|J_{AB}|$, as follows from Eqs. (5) and (12). This collapse of an AB pattern into an A_2 pattern is quite dramatic. Consider a trans-1,2-disubstituted ethylene with $J_{AB} = 15$ Hz and $\nu_A - \nu_B = 3$ Hz. The splitting b becomes $9/30 = 0.3$ Hz and will probably not be resolved. The intensity of one outer line is only $(3^2/4 \cdot 15^2)/2 \simeq 0.005$ of the intensity of the collapsed center doublet! It is evident that the width of the collapsed doublet can hardly be used for an estimate of the chemical shift difference $(\nu_A - \nu_B)$.

d) Summary

1. Basic Structure of an AB Spectrum

1.1. A two-spin AB system always produces a symmterical four-line pattern. The two outermost lines are separated by a frequency interval a, and the two inner lines by the interval b (Fig. 24). The intensity I_b of an outer line is related to that of an inner line I_a by the simple ratio:

$$I_a/I_b = a/b\,. \tag{13}$$

1.2. Even if the rules of paragraph A above are satisfied, ambiguities *may* still arise. Thus, if $a = 3b$, the AB basic multiplet displays a simple 1 : 3 : 3 : 1 pattern corresponding in total intensity to two spins. This is indistinguishable from the 1 : 3 : 3 : 1 pattern in the A_2 basic multiplet of an A_2X_3 spin system, as would be obtained, for example, from the methylene protons of 1,1,1-trifluoro-2-chloro-ethane.

1.3. The spectrum is naturally independent of our choice of labels A and B and we may always choose $\nu_A > \nu_B$. The spectrum is also independent of the sign of J_{AB}.

2. Step-by-step Analysis of an AB Spectrum

Step 1. Measure the centre frequency $\bar{\nu}$ and the frequency intervals a and b (Fig. 24).

Step 2. Evaluate the quantity

$$\Delta = \sqrt{ab}$$

and the NMR parameters according to

$$\nu_A = \bar{\nu} + \Delta/2$$
$$\nu_B = \bar{\nu} - \Delta/2$$
$$|J_{AB}| = (a-b)/2$$

This completes the AB analysis.

Example 1: The Proton Resonance Spectrum of 2,3-dibromothiophene

The proton resonance spectrum of the title compound (15% by weight in $CDCl_3$) recorded at 56.4 MHz is depicted in Fig. 25. The observed absorption lines occur at (in Hz) 410.39, 404.65, 390.64 and 384.87 downfield from TMS.

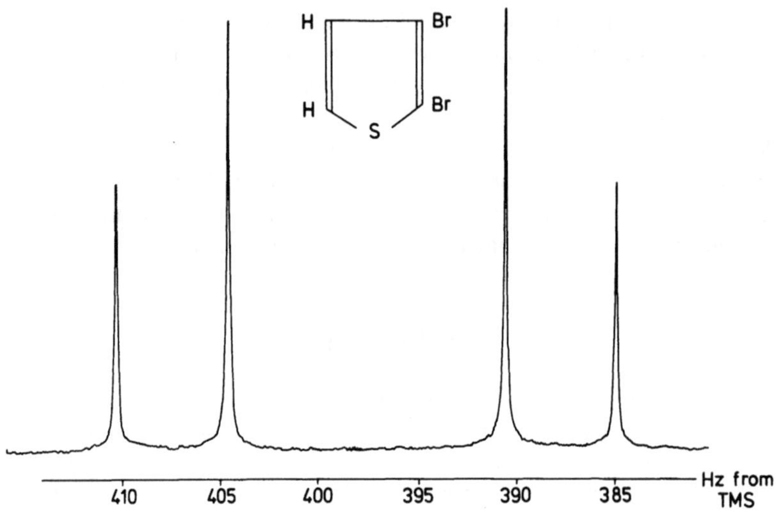

Fig. 25. NMR spectrum at 56.4 MHz of 2,3-dibromothiophene in a $CDCl_3$ solution (15% by weight)

This gives $a = 25.52$ Hz and $b = 14.01$ Hz and a centre frequency of 397.63 Hz (average of the two sets of line pairs). This gives a chemical shift difference $\Delta = \sqrt{ab} = 18.91$ Hz between the two protons and a coupling constant $J = (a-b)/2 = 5.70$ Hz. From the analysis, it is not possible assign the signals to a specific proton. However, from more detailed studies it has been established that the 5-proton is the more shielded of the two protons, and we can consequently make an assignment of chemical shifts to obtain $\nu_A = \nu_5 = 407.10$ Hz and $\nu_B = \nu_4 = 388.18$ Hz. The predicted intensity ratio is $I_a/I_b = a/b = 1.8$, which compares reasonably well with the experimental value of 1.65.

Example 2: The Proton Spectrum of Trans-1-bromo-2-chloroethene

An example of a more strongly coupled system is offered in the title compound, the spectrum of which is shown in Fig. 26. In this strongly coupled spectrum the outer lines are very weak and might even escape detection (under unfavourable noise conditions). Whereas a reasonable estimate of the chemical shift could have been obtained from the spectrum in the previous example by a simple first order analysis, such an attempt would lead to very large errors in the present example. The centre frequency is found at

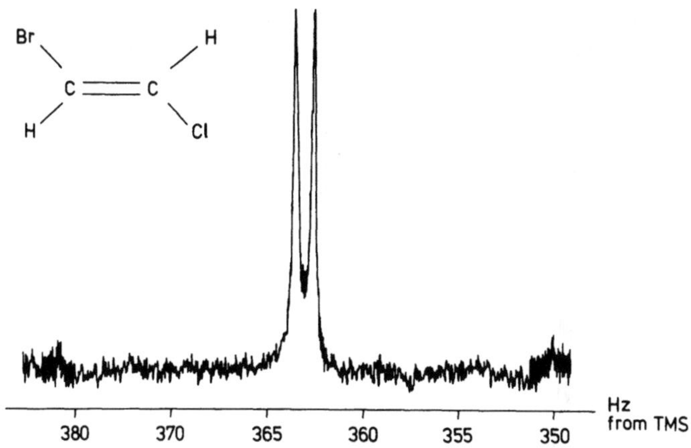

Fig. 26. Proton magnetic resonance spectrum at 56.4 MHz of trans-1-bromo-2-chloroethene in a hexafluorobenzene solution. The regions between the outer and inner peaks have been scanned at a faster sweep rate, which explains the different appearance of the noise in different regions

$\bar{\nu} = 363.16$ Hz downfield from TMS and the spacings $a = 25.80$ Hz and $b = 0.89$ Hz. This gives a chemical shift difference $\Delta = 4.76$ Hz and a spin-coupling constant $J = 12.46$ Hz. In the present case the predicted intensity ratio between the inner and outer lines $I_a/I_b = a/b = 29:1$. The observed intensity ratio is $20:1$. This discrepancy is due to a beginning saturation of two inner strong lines which, as discussed in part III, are easily saturated due to their large transition probability.

An assignment of the chemical shifts to the two protons cannot be safely made in this case.

IV.3. Deceptive Simplicity

a) Definition of Deceptive Simplicity

In Section IV.2 it was shown that an AB spectrum may have the appearance of an A_2 spectrum, even though the chemical shift $\nu_A - \nu_B$ is larger than the experimental line-widths. The reason for this deceptive simplicity of the AB spectrum was found to be that all structural features in the spectrum that depend on $(\nu_A - \nu_B)$ vanish with the *square* of $(\nu_A - \nu_B)$.

When features depending on the non-vanishing value of an NMR parameter disappear at a *higher* rate than the parameter itself vanishes, one is faced with the phenomenon of deceptive *simplicity.*

Deceptive simplicity may be encountered in a large variety of spin systems but is particularly common in systems containing symmetrically equivalent spins. Deceptive simplicity always arises through strong coupling of some nuclei among themselves.

Example 1: The Proton Resonance Spectrum of Furan

The 56.4 MHz proton magnetic resonance spectrum of furan is depicted in Fig. 27. On inspection, the spectrum may be divided into two basic multiplets of 1 : 2 : 1 triplets. The low field triplet is assigned to the α-protons and the high field

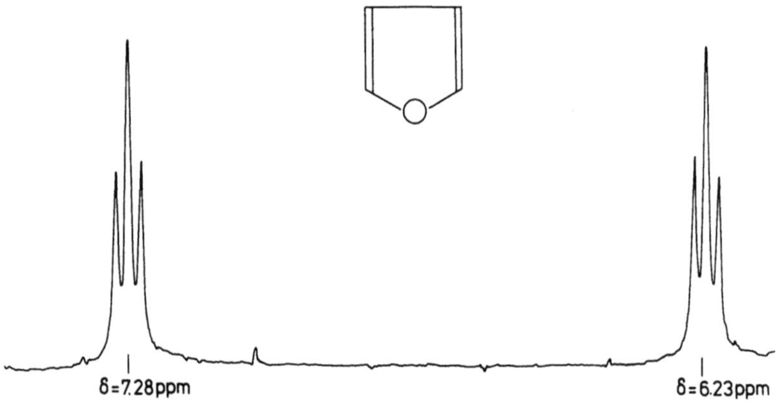

Fig. 27. NMR spectrum at 56.4 MHz of furan in a cyclohexane (30% by weight) solution. The small peaks between the two triplets are due to impurities

triplet to the β-protons. The structure of the multiplets could obviously be accounted for by a simple first order A_2X_2 analysis, in which the two α-protons are taken as one pair of magnetically equivalent nuclei and the β-protons as a second pair of magnetically equivalent nuclei. On this basis, the observed splittings in the triplets would represent the spin-coupling constant $J_{\alpha\beta}$ and the difference between J_{23} and J_{24} would be assumed to be negligible in order to satisfy the requirement of magnetic equivalence.

The only difficulty with this interpretation is the lack of any obvious reason for equality of J_{23} and J_{24}. If one refrains from this unwarranted assumption, the magnetic equivalence is removed and the simple first order analysis must be replaced with an AA′XX′ analysis. As will be shown in Section IV.9 the multiplet structure in an AA′XX′ spectrum is usually more complex than that of an A_2X_2 first order spectrum, but this difference depends on the non-vanishing value of the single parameter $L = J_{AX} - J_{A'X}$ (*i.e.* $J_{23} - J_{24}$ in furan) and vanishes quadratically with vanishing L. The spectrum of furan is therefore deceptively simple [*10*], and the observed splitting in the two triplets represents in fact the average value of the couplings between α and β protons, *i.e.* $(J_{23} + J_{24})/2 = (J_{45} + J_{35})/2$.

Example 2: The Proton and ^{19}F resonance Spectrum of 1,4-Difluorobenzene

The proton resonance spectrum of 1,4-difluorobenzene reproduced in Fig. 28a resembles that of a simple 1 : 2 : 1 triplet pattern and the ^{19}F resonance spectrum (Fig. 28b) resembles that of a 1 : 4 : 6 : 4 : 1 quintet. The total spectrum thus looks like a simple first order A_4X_2 system. The geometry of the molecule does not, however, suggest magnetic equivalence of the four protons or of the two fluorines, since the *ortho* and *meta* J_{HF} spin coupling constants would be expected to be unequal. As in the case of furan, the deviation from magnetic

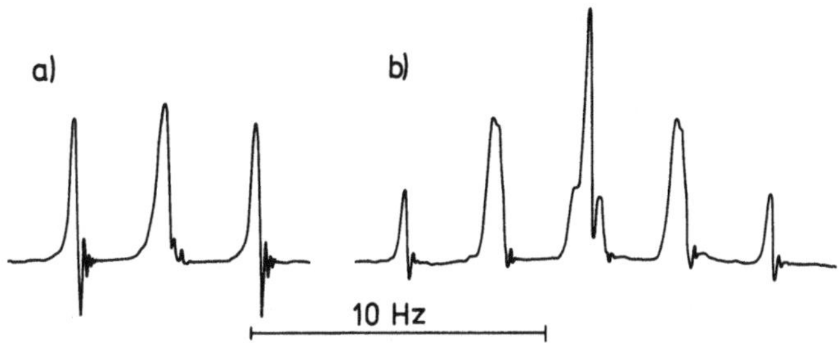

Fig. 28. The NMR spectrum of 1,4-difluorobenzene at 56.4 MHz. a The proton magnetic resonance region. b The fluorine (^{19}F) resonance spectrum. The proton magnetic resonance has been recorded with reduced gain

equivalence depends only on the difference of two coupling constants. The proton-proton coupling constants, as well as the fluorine-fluorine coupling constant, which tend to make the symmetrically equivalent nuclei act in unison, are apparently more effective in determining the structure of the spectrum than is the counteracting tendency of the difference $J_{HF}^{ortho} - J_{HF}^{meta}$. That the difference $J_{HF}^{ortho} - J_{HF}^{meta}$ is not completely negligible, however, can be seen from the spectra in Fig. 28 where some line-broadening effects are clearly discernable, showing the presence of additional structure in the lines. Some of this structure may be seen under improved resolution.

The observed multiplet splittings again correspond to the average of J_{HF}^{ortho} and J_{HF}^{meta}.

b) Apparent Magnetic Equivalence

A general rule may be formulated on the splitting patterns in spectra which are deceptively simple by apparent magnetic equivalence of spins that, for geometrical reasons, should not be magnetically equivalent: the splittings corresponding to couplings between the apparently magnetically equivalent spins and other spins represent the arithmetic mean of the actual coupling constants.

The apparently equivalent nuclei behave as if they shared equitably among themselves all their spin couplings to nuclei outside the group.

c) Some Aids in Overcoming Deceptive Simplicity

A number of methods may be employed to obtain additional information from deceptively simple spectra. The first method is conceptually simple but laborious from an experimental point of view because it requires the synthesis of isotope labelled compounds. Consider an AB spin system that has the deceptively simple appearance of an A_2 band assigned to two protons with a small relative chemical shift. If one of the protons is replaced with a deuteron, the spectrum automatically becomes first order. Apart from being laborious, this method suffers from the disadvantage that the spin coupling decreases in magnitude by a factor of $\gamma_D/\gamma_H \simeq 0.15$ and the accuracy is correspondingly reduced.

The second method, which involves the recording of *satellite* bands originating from transitions in molecules containing less abundant nuclei, such as ^{13}C and ^{29}Si, may be demanding on spectrometer performance since these bands are quite weak in non-enriched samples. Consider a molecule CHX_2-CHY_2 where X and Y are not spin-coupled to the two protons. If the chemical shift between the two protons is small compared to their spin-coupling constant, the spectrum of the most abundant species containing only ^{12}C nuclei will degenerate into a simple A_2 spectrum. However, molecules containing one ^{13}C nucleus will produce ABX spectra and these ABX spectra will not be deceptively simple owing to the large difference between J_{AX} and J_{BX}. The structure of the proton resonance spectrum (the AB basic multiplet) allows determination of both $|J_{AB}|$ and $|\nu_A-\nu_B|$. The latter parameter will probably not be very useful, however, because correlation tables are not reliable enough to allow a unique assignment of ν_A and ν_B to the two protons.

The third method, already discussed in part III, es experimentally very simple since it involves only a repetition of the NMR experiment at increased r.f. amplitude. In every new scan the r.f. amplitude is increased by a factor of the order of 1.5 (*ca.* 3 dB). The part of the spectrum to be recorded is centered on the collapsed AB line and should have a width corresponding to slightly more than twice the expected value of the spin-coupling constant J_{AB} in order to cover the interval a within which the two lines of low intensity are located. This procedure is repeated until these two lines are found. Once the weak lines have been observed the frequency interval a in Fig. 24 is known. Since the frequency interval b is unobservably small, it follows from Eq. (4) that J_{AB} is only slightly less than $a/2$. Since b has not been measured, the chemical shift $\nu_A-\nu_B$ cannot be evaluated from Eq. (3) but the width of the collapsed center line enables one to give an upper limit for the quantity b and consequently to $|\nu_A-\nu_B|$.

If the two weak lines have not been found when the r.f. amplitude has been increased by a factor of, say 25, over that normally employed in studying dilute samples (giving an optimum signal-to-noise ratio of 40 : 1 of the unsplit band of a single proton), this method of increasing the information content will probably not succeed and can be abandoned.

It should be noted that this phenomenon of selective saturation may be employed to distinguish between lines that are weak owing to small transition probabilities from lines that are weak because of low concentration, such as impurity bands, and lines caused by spin coupling to less abundant nuclei like ^{13}C, ^{29}Si etc.

IV.4. The Three-Spin System ABX

a) The Energy-Level Diagram of an ABX Spin System

The ABX spin system is sometimes defined as a system of three spins 1/2 in which the spin coupling J_{AB} is comparable in magnitude with the chemical shift difference $\nu_A - \nu_B$, whereas J_{AX} and J_{BX} are small in magnitude in comparison with both $\nu_A - \nu_X$ and $\nu_B - \nu_X$. For reasons that will become obvious in the examples discussed below, we prefer to define the ABX spin system as a three-spin system with two basic groups AB and X.

The labelling of stationary states and transitions in an ABX spin system is usually modelled on the correspondence with the weakly coupled APX spin system. The energy-level and transition diagram of the APX spin system is given in Fig. 29.

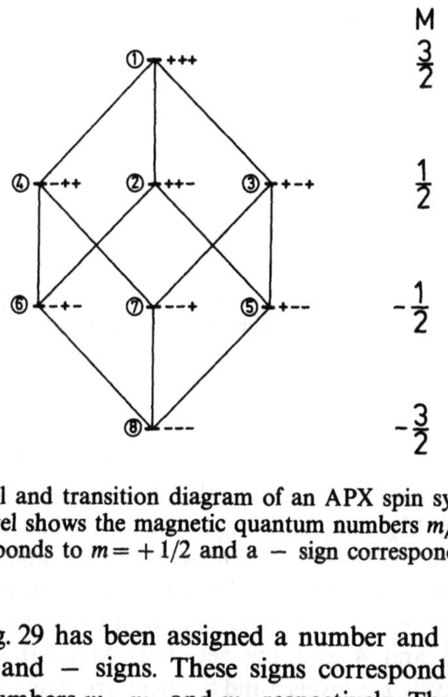

Fig. 29. The energy level and transition diagram of an APX spin system. The sequence of $\pm$ signs next to each level shows the magnetic quantum numbers m_A, m_P, m_X in each state, a + sign corresponds to $m = +1/2$ and a $-$ sign corresponds to $m = -1/2$

Every level in Fig. 29 has been assigned a number and a symbol consisting of a sequence of + and $-$ signs. These signs correspond to the signs of the magnetic quantum numbers m_A, m_P and m_X respectively. Thus $+ - +$ represents a state in which $m_A = +1/2$, $m_P = -1/2$ and $m_X = +1/2$. The ordering of the states in the diagram may seem irrational but is caused by a desire to keep the numbering in conformity with that of standard texts while at the same time preparing for the analysis of the AB basic multiplet of an ABX spin system. In Fig. 29 the X transitions, *i.e.* transitions in which m_X changes by unity, are represented by vertical lines, the A transitions are represented by lines slanting downwards to the left and the P transitions by lines slanting downwards to the right. To the right of the energy-level and transition diagram is indicated the value of the total magnetic quantum number M of the spin system.

The value of the total magnetic quantum number M is retained irrespective of the extent of coupling in the spin system, and the ABX spin system will also have one level with M = + 3/2, three with M = + 1/2, three with M = − 1/2 and one with M = − 3/2. Thus this characteristic is retained in the ABX transition and energy-level diagram depicted in Fig. 30. However, the individual magnetic quantum numbers m_A and m_B are no longer characteristic of a stationary state in an ABX spin system, and consequently a label such as + − + is no longer appropriate for level no. 3.

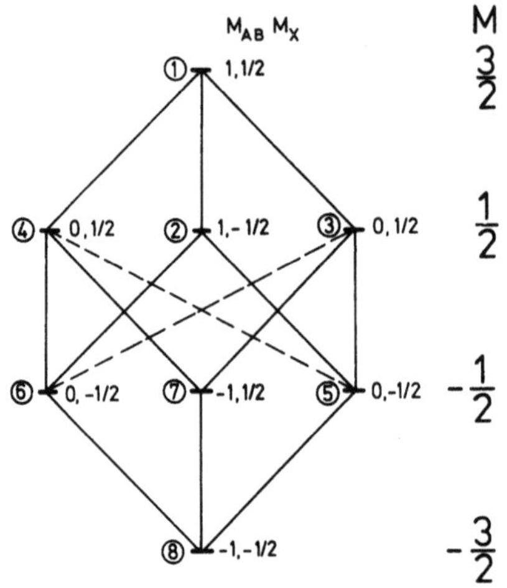

Fig. 30. The energy-level and transition diagram of an ABX spin system. The set of numbers next to each energy level shows the values of the magnetic quantum numbers M_{AB} and M_X in each state. The broken lines are X transitions which become forbidden in the APX limit

In addition to the total magnetic quantum number M, the energy levels of the ABX spin system are also characterized by the magnetic quantum numbers of the basic groups viz. $M_X = m_X$ and $M_{AB} = M - m_X$. The observed lines in the spectrum correspond to transitions in which one, and only one, of these magnetic quantum numbers changes by unity. The set of magnetic quantum numbers $M_{AB} = 0$ and $m_X = +1/2$ corresponds to two levels in the ABX spin system, viz. levels 3 and 4 and the set with $M_{AB} = 0$ and $m_X = -1/2$ also corresponds to two levels, *viz.* levels 5 and 6. Therefore levels 3 and 4 in an ABX spin system belong to states obtained by mixing the states + − + and − + + of the corresponding levels of an APX spin system, and the states of levels 5 and 6 similarly represent mixed states, but the states 1, 2, 7 and 8 are unmodified from the weakly coupled APX spin system.

The observed lines in the ABX spectrum correspond to transitions in which one and only one of the magnetic quantum numbers M_{AB} or M_X changes by unity. The X basic multiplet consists of transitions in which M_X changes by unity and M_{AB} remains unchanged. Thus there are 6 possible transitions in the X basic multiplet, viz. 1→2, 3→5, 3→6, 4→5, 4→6, and 7→8. The transitions 1→2 and 7→8 which occur between unmixed X states retain their first order character and occur at the frequencies

$$\nu_X + J_{AX}/2 + J_{BX}/2$$

and

$$\nu_X - J_{AX}/2 - J_{BX}/2$$

respectively.

As in the first order case each of these lines has an intensity that is 1/4 of the total intensity of the X basic multiplet. The remaining half of the X basic multiplet intensity is shared among the four remaining X transitions in a manner that depends on the mixing of the AB spin states. However, none of these four transitions can become stronger than the two first order transitions in the X multiplet. The transitions $3 \rightarrow 6$ and $4 \rightarrow 5$, which have no counterpart in the APX spin system, are sometimes called combination lines.

Every transition in the AB basic multiplet involves one mixed state and, in general, no unmodified first order lines are to be found in the AB basic multiplet. Therefore it is seldom helpful to analyse the AB basic multiplet by making correspondence with the first order APX spectrum. Instead, the AB basic multiplet should be divided into two subspectra. One AB *subspectrum* involves the four transitions $1 \rightarrow 3$, $1 \rightarrow 4$, $3 \rightarrow 7$ and $4 \rightarrow 7$ arising in molecules *which have* $m_X = +1/2$, and *the other* subspectrum involves the transitions $2 \rightarrow 5$, $2 \rightarrow 6$, $5 \rightarrow 8$ and $6 \rightarrow 8$ arising in molecules *which have* $m_X = -1/2$. The total intensity of the AB basic multiplet is shared evenly by the two subspectra because molecules with $m_X = -1/2$ are just as frequent (usually to within a few parts in 10^5) as are molecules with $m_X = +1/2$. From Fig. 30 it is evident that the transitions within one such subspectrum form a closed quadrangle with an unmixed state at the top and one at the bottom and two mixed states at an intermediate energy. This is exactly the same pattern as that of a two-spin AB system discussed in Section IV.2. The full quantum mechanical analysis of the ABX spin system reveals that the four-line pattern of each *one* of these subspectra displays a mirror-image symmetry and satisfies the same relationship between transition frequencies and relative intensities as that of an actual two-spin AB system (cf. Eq. (6)). These patterns are therefore known as *AB* or *ab* subspectra; lower case letters (*ab*) may be used to emphasize that the apparent shifts and coupling constants of a subspectrum need not equal those of the actual spin system. From the full treatment of ABX in Section V.8 it follows in fact that the apparent *ab* spin coupling constants of the two subspectra are equal to one another and both equal to J_{AB}, the actual A–B spin-coupling constant of our ABX system. On the other hand, the apparent Larmor frequencies ν_a and ν_b of one AB-subspectrum in ABX are usually different from that of the other and, in general, none of them equals ν_A or ν_B. The apparent and real Larmor frequencies are, however, simply related to one another in the present case by the concept of effective Larmor frequency.

b) The Concept of Effective Larmor Frequencies

The concept of effective Larmor frequencies is best introduced with reference to the simple weakly coupled two-spin system AX. In the AX system we have only two Larmor frequencies according to the definition of Eq. (10) of Section I.2. The four transition frequencies observed in the AX spin system may, however, be viewed as four *effective* Larmor frequencies for the following reasons.

The interaction between spin A and the H_0 field as modified by screening is described by the equation

$$E = h\nu_A I_z(A) \tag{14}$$

and the coefficient of $I_z(A)$ contains, apart from Planck's constant h, a constant of dimension frequency, *viz.* ν_A. The spin-spin coupling interaction in the weakly coupled AX spin system may be written

$$E = hJ_{AX} I_z(A)\, I_z(X) \tag{15}$$

with neglect of the transverse components of the spin vectors. In transitions between spin states of nucleus A, the component $I_z(X)$ remains constant and has the value of $m_X(+1/2$ *or* $-1/2)$.

Thus the interaction of Eq. (15), when only A transitions are considered, may be rewritten

$$E = hJ_{AX} I_z(A)\, m_X\,. \tag{16}$$

Combining Eqs. (14) and (16) we obtain

$$E = h(\nu_A + J_{AX} m_X)\, I_z(A) \tag{17}$$

and the frequency $\nu_A + J_{AX} m_X$ has the character of an *effective Larmor frequency* for spin A in molecules where $I_z(X) = m_X$. In a similar manner we obtain a pair of effective Larmor frequencies for nucleus X, *i.e.* $\nu_X - J_{AX}/2$ and $\nu_X + J_{AX}/2$, corresponding to the two possible values of m_A.

c) Effective Larmor Frequencies in the ABX Spin System

In the AX spin system the effective Larmor frequencies correspond to actual transition frequencies in the spectrum, but this is usually not the case in the AB subspectra of an ABX spin system.

The A and B spins are strongly coupled and therefore the J_{AB} spin-spin interaction cannot be disposed of by the effective Larmor frequency approach. In each of the two subspectra of the AB basic multiplet, however, the weakly coupled nucleus X has a spin vector that is described entirely by its magnetic quantum number m_X and therefore the spin-spin interactions J_{AX} and J_{BX} can be incorporated into effective Larmor frequencies. In molecules where

$$m_X = +1/2$$

we obtain the effective Larmor frequencies ν_a^+ and ν_b^+

$$\nu_a^+ = \nu_A + J_{AX}/2\,, \tag{18}$$

$$\nu_b^+ = \nu_B + J_{BX}/2\,, \tag{19}$$

and in molecules with

$$m_X = -1/2$$

we obtain the effective Larmor frequencies

$$\nu_a^- = \nu_A - J_{AX}/2\,, \tag{20}$$

$$\nu_b^- = \nu_B - J_{BX}/2\,. \tag{21}$$

The pairs of effective Larmor frequencies (18) and (19) or (20) and (21) and the spin-coupling constant J_{AB} are the parameters that determine the structure of the two subspectra in the AB basic multiplet.

From Eqs. (18)–(21) one obtains the shift

$$\nu_A - \nu_B = (\nu_a^+ - \nu_b^+)/2 + (\nu_a^- - \nu_b^-)/2$$

while $(J_{AX} - J_{BX})/2$ is given by the expression

$$(J_{AX} - J_{BX})/2 = (\nu_a^+ - \nu_b^+)/2 - (\nu_a^- - \nu_b^-).$$

When the two ab-subspectra are analysed in the manner of Section IV.2.b, however, one will obtain only the *magnitudes* of $|\nu_a^+ - \nu_b^+|$ and $|\nu_a^- - \nu_b^-|$ but not their *signs*. One of the consequences of this sign ambiguity is that one will not be able to distinguish between $|\nu_A - \nu_B|$ and $|J_{AX} - J_{BX}|/2$, so that the chemical shift $|\nu_A - \nu_B|$ of an ABX spectrum cannot be obtained by analysing solely the AB basic multiplet!

d) Analysis of Observed ABX Spectra

The problem of analysing an experimental spectrum has two aspects. First, one has to classify the observed lines according to how they originate in the transition and energy-level diagram. This is a problem of spectra assignment which may have to be solved for each individual line, or, as in the present case, for certain groups of lines known as subspectra. The second aspect involves the extraction of the NMR parameters from the observed transition frequencies and intensitites of the lines so assigned.

In part, the first problem may be solved simply by studying the energy-level diagram for a general case to derive frequency sum rules etc. The second problem, on the other hand, usually necessitates a quantum mechanical calculation. In the spin systems discussed in the present chapter the quantum mechanical calculation results in analytical relationships in closed form and it therefore becomes possible to prescribe a step-by-step procedure that will yield the desired result.

Before describing such a step-by-step analysis procedure for an ABX spin system we must, however, summarize the basic features of an ABX spectrum as derived in part from the foregoing discussion and in part from the full quantum mechanical treatment.

e) Basic Structure of an ABX Spectrum

(i) The spectrum of a genuine ABX spin system displays two distinct bands, *viz.* a symmetrical six-line pattern in the X basic multiplet and an eight-line pattern in the AB basic multiplet. Noise and overlap may reduce the number of actually observed lines in an experimental ABX spectrum.

The centre of *symmetry* of the X basis multiplet corresponds to ν_X. (The centre of *gravity* of the AB basic multiplet corresponds to $(\nu_A + \nu_B)/2$, but ν_A and ν_B are better evaluated in the step-by-step analysis.)

(ii) The AB basic multiplet may always be decomposed into two subspectra and the total intensity of the multiplet is shared evenly by the two subspectra.

The four-line pattern of each *one* of the two subspectra displays a mirror-image symmetry and has the same intensity ratio between inner and outer lines as that of a two-spin AB spectrum. The apparent A–B coupling constants of the two subspectra are equal and both given by J_{AB}, the actual A–B spin-coupling constant of our ABX spin system.

(iii) The AB basic multiplet is invariant under an interchange of the magnitudes of $|\nu_A - \nu_B|$ and $|J_{AX} - J_{BX}|/2$. The transition *frequencies* of the X basic multiplet are also invariant under such an exchange, but their relative *intensities* are not.

(iv) The spectrum remains unchanged if we change the signs of J_{AX} and J_{BX} simultaneously and also if we change the sign of J_{AB}. We shall *arbitrarily* choose an analysis procedure that always yields positive values for J_{AB} and for the larger one of J_{AX} and J_{BX}. The *relative* signs of J_{AX} and J_{BX} are usually obtained through the analysis.

f) Step-by-step Analysis of an ABX Spectrum

In our procedure *labels* A and B are always chosen to yield $\nu_A > \nu_B$.

Step 1. Pick out the two subspectra of the AB basic multiplet by use of the rules under e.(ii). above.

Step 2. Measure the centre frequencies $\bar{\nu}_1$ and $\bar{\nu}_2$ and the frequency intervals a_1, b_1 and a_2, b_2 of the two subspectra (Fig. 31). Always choose labels so that $\bar{\nu}_1 > \bar{\nu}_2$.

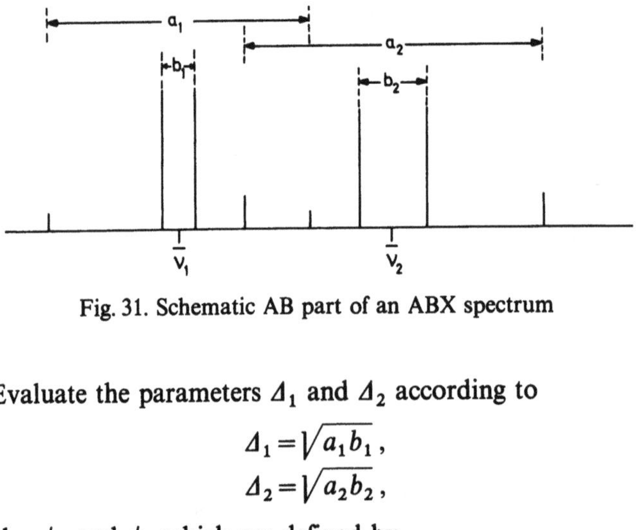

Fig. 31. Schematic AB part of an ABX spectrum

Step 3. Evaluate the parameters Δ_1 and Δ_2 according to

$$\Delta_1 = \sqrt{a_1 b_1}, \quad (22)$$

$$\Delta_2 = \sqrt{a_2 b_2}, \quad (23)$$

and the angles ϕ_1 and ϕ_2 which are defined by

$$\sin 2\phi_1 = \frac{a_1 - b_1}{a_1 + b_1}, \quad (24)$$

$$\sin 2\phi_2 = \frac{a_2 - b_2}{a_2 + b_2}, \quad (25)$$

and chosen in the interval $0° < \phi < 45°$.

Step 4. Prior to a comparison of calculated and observed X multiplet intensity distributions, we must, according to e.(iii). above, in principle accept two different sets of NMR parameters. In set (*i*) we assume that $|\nu_A - \nu_B| > |J_{AX} - J_{BX}|/2$ and in set (*ii*) this inequality is reversed. The parameters of set (*i*) are given by:

Set (i).

$$J_{AB} = (a_1 - b_1)/2 = (a_2 - b_2)/2\,, \tag{26}$$

$$\nu_A = (\bar{\nu}_1 + \bar{\nu}_2)/2 + (\Delta_1 + \Delta_2)/4\,, \tag{27}$$

$$\nu_B = (\bar{\nu}_1 + \bar{\nu}_2)/2 - (\Delta_1 + \Delta_2)/4\,, \tag{28}$$

$$J_{AX} = (\bar{\nu}_1 - \bar{\nu}_2) + (\Delta_1 - \Delta_2)/2\,, \tag{29}$$

$$J_{BX} = (\bar{\nu}_1 - \bar{\nu}_2) - (\Delta_1 - \Delta_2)/2\,, \tag{30}$$

and ν_X is given by the centre of symmetry of the X basic multiplet.

In set (*ii*) the NMR parameters are given by the same set of equations with the sign of the *smaller* one of Δ_1 and Δ_2 reversed in Eqs. (27) to (30).

Step 5. Because of the existence of frequency and intensity sum rules, it is usually convenient to analyse the X basic multiplet in terms of the same entities as those derived in an AB basic multiplet. When X is of a different isotopic species, however, it often happens that the magnetic field H_0' employed in recording the X basic multiplet is different from the field H_0 employed for recording the AB basic multiplet. If $H_0'/H_0 = 1$ one may proceed directly to step 6 below. If on the other hand $H_0'/H_0 \neq 1$ the value of $(\nu_A - \nu_B)$ at H_0' differs from the value at H_0 and the entities $(a_1 + b_1)$ and $(a_2 + b_2)$ occurring in Eqs. (24)–(25) above as well as in Eqs. (34)–(36) below must first be modified according to the formulas:

$$a_1' + b_1' = 2\sqrt{J_{AB}^2 + \frac{1}{4}\left\{\frac{H_0'}{H_0}(\Delta_1 + \Delta_2) + (\Delta_1 - \Delta_2)\right\}^2} \tag{31}$$

$$a_2' + b_2' = 2\sqrt{J_{AB}^2 + \frac{1}{4}\left\{\frac{H_0'}{H_0}(\Delta_1 + \Delta_2) - (\Delta_1 - \Delta_2)\right\}^2} \tag{32}$$

whereas $(a_1' - b_1')$ and $(a_2' - b_2')$ retain their value of $2J_{AB}$:

$$a_1' - b_1' = a_2' - b_2' = 2J_{AB}\,. \tag{33}$$

Step 6. In order to compare the calculated and observed intensities of the three pairs of X multiplet lines centered at ν_X, we shall first normalize the total intensity of the X basic multiplet to 4 units.

One pair of lines, that will always be strong, is separated by a frequency interval of

$$|J_{AX} + J_{BX}| = 2|\bar{\nu}_1 - \bar{\nu}_2| \tag{34}$$

and each one of these transitions has an intensity of unity. A second pair displays an interval of

$$|a_1 + b_1 - a_2 - b_2|/2 \tag{35}$$

and the intensity of each line equals $\cos^2(\phi_1 - \phi_2)$ if set (*i*) is correct, but is equal to $\sin^2(\phi_1 + \phi_2)$ if set (*ii*) is the correct one. The third pair displays a larger splitting of

$$|a_1 + b_1 + a_2 + b_2|/2 \tag{36}$$

and the intensity of each line must equal $\sin^2(\phi_1 - \phi_2)$ if set (*i*) is correct, but equals $\cos^2(\phi_1 + \phi_2)$ if set (*ii*) is correct.

g) Analysis of an Artificial ABX Spectrum

The spectrum may be though of as arising from a molecule containing two protons and one fluorine atom. The "observed" parameters taken at the same magnetic field H_0 for both ^1H and ^{19}F indicated in Fig. 32 are

$$\bar{\nu}_1 = 360.00\ \text{Hz}\,,$$
$$\bar{\nu}_2 = 349.00\ \text{Hz}$$

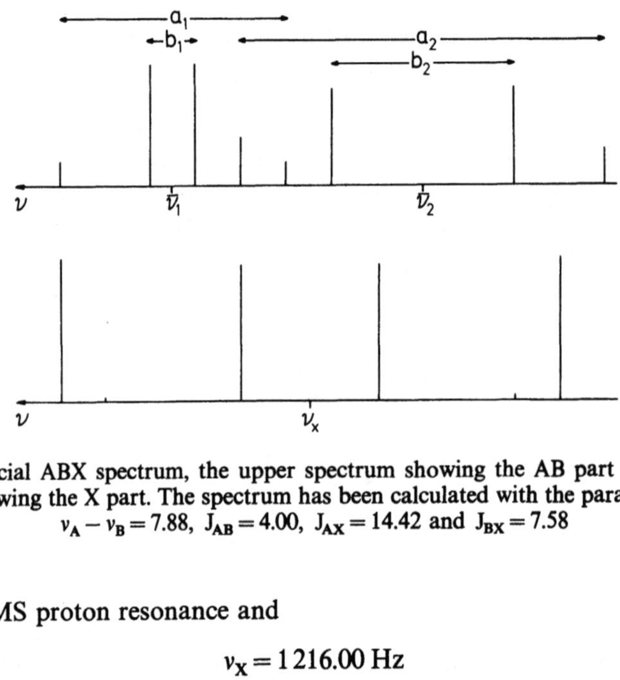

Fig. 32. Artificial ABX spectrum, the upper spectrum showing the AB part and the lower spectrum showing the X part. The spectrum has been calculated with the parameters (in Hz) $\nu_A - \nu_B = 7.88$, $J_{AB} = 4.00$, $J_{AX} = 14.42$ and $J_{BX} = 7.58$

from the TMS proton resonance and

$$\nu_X = 1216.00\ \text{Hz}$$

from the CCl_3F fluorine reference signal, and the frequency intervals are given by

$$a_1 = 10.00\ \text{Hz}\,,$$
$$b_1 = 2.00\ \text{Hz}\,,$$
$$a_2 = 16.00\ \text{Hz}\,,$$
$$b_2 = 8.00\ \text{Hz}\,.$$

From Eq. (26) we immediately obtain that

$$J_{AB} = 4.00\ \text{Hz}$$

and from Eq. (34) we obtain for the magnitude of $(J_{AX}+J_{BX})$

$$|J_{AX}+J_{BX}|=2|\bar{\nu}_1-\bar{\nu}_2|=22.00\ \text{Hz}.$$

The frequency of 22.00 Hz corresponds to the separation between lines X2 and X5 in the X basic multiplet and consequently these two lines are the two "first order" lines with theoretical intensities equal to 1/4 of the total intensity of the X multiplet. The two remaining strong lines X3 and X4 in the X multiplet must be assigned to "mixed" transitions.

From Eqs. (22) and (23) we obtain

$$\Delta_1=\sqrt{a_1 b_1}=4.472\ \text{Hz},$$
$$\Delta_2=\sqrt{a_2 b_2}=11.314\ \text{Hz}.$$

Now $\Delta_2>\Delta_1$, and we should therefore evaluate Eqs. (27) to (30) with $\Delta_2>0$. By choosing $\Delta_1>0$ as well, we obtain as a first set, set (*i*), of NMR parameters

(*i*)
$$\begin{aligned}\nu_A &= 358.44\ \text{Hz} \quad \text{rel. to TMS}\\ \nu_B &= 350.56\ \text{Hz} \quad \text{rel. to TMS}\\ J_{AX} &= 14.42\ \text{Hz}\\ J_{BX} &= 7.58\ \text{Hz}\\ J_{AB} &= 4.00\ \text{Hz}\\ \nu_X &= 1216.00\ \text{Hz} \quad \text{rel. to } CCl_3F.\end{aligned} \tag{37}$$

If this parameter set is the correct one, then the intensities of lines X1 and X6 must be given by $\sin^2(\phi_1-\phi_2)$ and the intensities of lines X3 and X4 by $\cos^2(\phi_1-\phi_2)$ with angles ϕ_1 and ϕ_2 determined by Eqs. (24) and (25). These latter equations yield

$$\phi_1=20\overset{\circ}{.}90,$$
$$\phi_2=9\overset{\circ}{.}73,$$

and one obtains the values

$$\begin{aligned}\cos^2(\phi_1-\phi_2)&=0.9625\\ \sin^2(\phi_1-\phi_2)&=0.0375.\end{aligned} \tag{38}$$

The relative intensities in set (*ii*) are given by

$$\begin{aligned}\sin^2(\phi_1+\phi_2)&=0.275\\ \cos^2(\phi_1+\phi_2)&=0.725.\end{aligned} \tag{39}$$

The "observed" X basic multiplet spectrum of Fig. 32 is evidently in good agreement with set (*i*), Eq. (38), since lines X3 and X4 are almost of unit intensity but not with set (*ii*) which predicts an intensity of lines X3 or X4 which is only 0.275 of that of line X2 or X5.

If the observed intensity distribution of the X basic multiplet had instead been that of Eq. (39), corresponding to set (*ii*), the NMR parameters of Eqs. (37) would have been wrong and the correct set would been obtained by evaluating Eqs. (27)

to (30) with a negative sign of Δ_1 yielding

(*ii*)

$$\begin{aligned} \nu_A &= 356.21\ \text{Hz} \quad \text{rel. to TMS} \\ \nu_B &= 352.79\ \text{Hz} \quad \text{rel. to TMS} \\ J_{AX} &= 3.11\ \text{Hz} \\ J_{BX} &= 18.89\ \text{Hz} \\ J_{AB} &= 4.00\ \text{Hz} \\ \nu_X &= 1216.00\ \text{Hz} \quad \text{rel. to } CCl_3F\,. \end{aligned} \tag{40}$$

It will be noted that the parameter set of Eq. (40) is quite different from that of Eq. (37) and is *not* obtained simply by changing the sign of some coupling constant.

h) Degeneracies in ABX Spectra

It is usually convenient to begin the analysis of an ABX spectrum by a decomposition of the AB basic multiplet into its two AB subspectra and the extraction of the NMR parameters according to Eqs. (26) to (30). This approach is simple when all the eight lines of the AB basic multiplet are resolved. It may happen, however, that fewer lines are actually observed.

First, one may encounter spectra in which only two of the eight transitions coincide in frequency to yield a seven-line AB basic multiplet, but since the splitting J_{AB} recurs four times in the AB basic multiplet, an accidental coincidence of *two pairs* of transition frequencies is not uncommon. The AB basic multiplet then degenerates into a six-line spectrum. In general, it should still be possible to pick out the two AB subspectra even in these degenerate cases.

When the magnitude of

$$|\nu_A - \nu_B| - |J_{AX} - J_{BX}|/2$$

is smaller than that of J_{AB}, one of the effective chemical shift differences, Δ_1 or Δ_2, of the AB subspectra becomes small enough to transform the corresponding AB subspectrum into a deceptively simple A_2-type subspectrum, consisting of only one strong line centered at $\bar{\nu}_1$ or $\bar{\nu}_2$ respectively (cf. Section IV.2.c above). The AB basic multiplet then degenerates into the AB-type subspectrum, with a total intensity equal to half the integrated intensity of the entire AB basic multiplet, and one single strong line comprising the second half of the total integrated intensity. One such spectrum is discussed in Example 2 below.

None of the two foregoing cases of degeneracy in the AB basic multiplet need correspond to any degeneracy in the X basic multiplet, and it is normally possible to extract all NMR parameters from such spectra.

The strongest degeneracy of an ABX spectrum is the collapse into a deceptively simple A_2X type spectrum. This will occur when the two effective chemical shift differences Δ_1 and Δ_2 are both small in comparison with J_{AB}. This, in turn, requires that both $|\nu_A - \nu_B|$ and $|J_{AX} - J_{BX}|/2$ be small in comparison with J_{AB}. Each one of the two AB subspectra then degenerates into a single line of half the total AB intensity and centered at $\bar{\nu}_1$ and $\bar{\nu}_2$ respectively. The midpoint between these two lines corresponds to $(\nu_A + \nu_B)/2$ and the separation between the lines equals $|J_{AX}+J_{BX}|/2$. The sum $(J_{AX}+J_{BX})$ is the algebraic sum of the two coupling constants and its magnitude is equal to the sum of $|J_{AX}|$ and $|J_{BX}|$ if J_{AX} and J_{BX} have the same signs, and equal to the difference between $|J_{AX}|$ and $|J_{BX}|$ if J_{AX} and J_{BX} have opposite signs. In the deceptively simple A_2X limit the X basic multiplet consists of a 1:2:1 triplet centered at ν_X and with splittings equal to $|J_{AX}+J_{BX}|/2$. The only NMR parameters which can be obtained in this case are $(\nu_A+\nu_B)/2$, ν_X and $|J_{AX}+J_{BX}|/2$. The two A and B nuclei are apparently magnetically equivalent (cf. Subsection b of Section IV.3).

The information obtainable from an X basic multiplet in an ABX spectrum is maximum when all six X lines can be observed. Frequently, however, fewer lines are recorded and the conditions for the disappearance or collapse of X multiplet lines depend primarily on the value of the parameter $|J_{AX}-J_{BX}|$.

Two of the six X lines are always very weak in intensity if $|J_{AX}-J_{BX}|<2|J_{AB}|$ while at the same time $|J_{AB}|<|\nu_A-\nu_B|$. If the chemical shift difference is smaller, so that $|J_{AB}|>|\nu_A-\nu_B|$, the condition for negligible intensity of two lines becomes more stringent, but as a role of thumb one may stipulate $|J_{AX}-J_{BX}|<|J_{AB}|/2$. An example of a spectrum where the former set of conditions is fulfilled is that of Fig. 33.

Collapse of the X basic multiplet into a 1:2:1 triplet occurs either because the spectrum is of the deceptively simple A_2X type or because

$$J_{AX}\simeq J_{BX} \tag{41}$$

with the two spin couplings of the same sign. The condition (41) is reasonably critical as long as $|\nu_A-\nu_B|\geqq|J_{AB}|$ because as long as this condition is fulfilled the splitting (Eq. (35)), which collapses to zero when $J_{AX}=J_{BX}$, remains larger than $0.7\,|J_{AX}-J_{BX}|$.

On rare occasions, one may encounter an X basic multiplet containing only two lines of equal intensity and separated by a splitting practically equal to $|J_{AX}+J_{BX}|$. This can only occur if J_{AX} and J_{BX} carry opposite signs, and the range of chemical shift values $|\nu_A-\nu_B|$ over which a two line X basic multiplet is obtained is of the order of the line-widths in the spectrum.

Finally, one may encounter ABX systems where $\nu_A=\nu_B$. In this case the two effective chemical shift parameters Δ_1 and Δ_2 must be equal in magnitude and enter Eqs. (27) to (30) with opposite signs, as follows from Eqs. (27) and (28). From Eqs. (29) and (30) it then follows that $|\Delta_1|=|\Delta_2|=|J_{AX}-J_{BX}|/2$. The two AB subspectra in the AB basic multiplet must therefore be identical excrpt for a translation of their centers equal to $\bar{\nu}_1-\bar{\nu}_2=|J_{AX}+J_{BX}|/2$. This is exactly the same AB basic multiplet as would have been obtained from a hypothetical second spin system A'B'X' in which $J_{A'X'}=J_{B'X'}$ and $|\nu_{A'}-\nu_{B'}|=|J_{AX}-J_{BX}|/2$. However, the line structure in the X basic multiplet will be quite different in the two cases. In the A'B'X' system the X' spectrum will display the usual 1:2:1 triplet with splittings equal to $|J_{A'X'}+J_{B'X'}|$. In the case at hand, we shall also have a triplet centered around ν_X and with a splitting of $|J_{AX}+J_{BX}|$ because the two X lines that are separated by $|a_1+b_1-a_2-b_2|/2$ coincide at ν_X. However, the intensity of this collapsed line will in general, *i.e.* as long as $|J_{AX}-J_{BX}|$ is sufficiently large to ensure that we have not reached the deceptively simple A_2X case, be less than half the total X intensity and two additional lines separated by a frequency of

$|a_1 + b_1 + a_2 + b_2|/2 = \sqrt{(J_{AX} - J_{BX})^2 + 4J_{AB}^2}$ will also be seen, thus giving a five-line X part spectrum. In fact, if

$$\nu_A = \nu_B$$

and

$$|J_{AX} - J_{BX}| \gg |J_{AB}|$$

the intensity of the line centered at ν_X vanishes and the whole ABX spectrum assumes a simple first order appearance.

Example 1. The Proton Resonance Spectrum of 2-methylthiofuran

The proton resonance spectrum of 2-methylthiofuran in a cyclohexane solution (25% by weight) is depicted in Fig. 33. Strictly, this spectrum is of the ABP_3X type, the P band being obtained from the thiomethyl group. The thio-

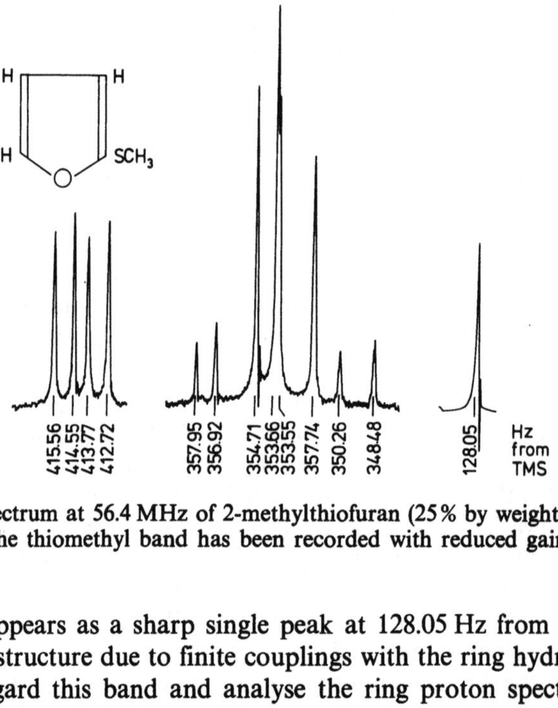

Fig. 33. NMR spectrum at 56.4 MHz of 2-methylthiofuran (25% by weight in cyclohexane). The thiomethyl band has been recorded with reduced gain

methyl band appears as a sharp single peak at 128.05 Hz from TMS, showing no sign of fine structure due to finite couplings with the ring hydrogens. We can therefore disregard this band and analyse the ring proton spectrum using the ABX step-by-step analysis scheme. The two *ab* subspectra in the AB part of the spectrum are readily recognized and are composed of the transitions at (in Hz) 357.95, 354.71, 353.55, 350.26 and 356.92, 353.66, 351.74, 348.48 respectively. The two central lines in the second (high-field) *ab* submultiplet (353.66 Hz and 351.74 Hz) do not show equal intensity in the spectrum in Fig. 33 due to overlap of the two transitions at 353.6 Hz. The assignment in this region has therefore been based on the line position data, since the splitting $|J_{AB}|$ should recur four times.

The centre frequencies of the two *ab* subspectra are given by $\bar{\nu}_1 = 354.13$ Hz and $\bar{\nu}_2 = 352.70$ Hz respectively. The relevant frequency intervals are given by (in Hz) $a_1 = 7.69$, $b_1 = 1.16$, $a_2 = 8.44$ and $b_2 = 1.92$. This gives the parameters

$\Delta_1 = 2.99$ Hz and $\Delta_2 = 4.03$ Hz. If we first evaluate the shift and coupling parameters according to the prescription for set (*i*) in Section IV.4.f above, *i.e.* with both Δ_1 and Δ_2 chosen positive, we get (in Hz) $J_{AB} = 3.26$, $\nu_A = 355.16$, $\nu_B = 351.66$, $J_{AX} = 0.91$, $J_{BX} = 1.95$. The centre of the X band gives us the shift $\nu_X = 414.15$ Hz downfield from TMS.

If this parameter set is correct, the X part of the spectrum should consist of one line pair of unit intensity separated by a frequency interval of 2.86 Hz, and one line pair of intensity 0.994 separated by 0.76 Hz. The third line pair of separation 9.60 Hz should have a relative intensity, of 0.006 and consequently be unobservable in the present spectrum.

This calculated spectrum agrees very well with the observed spectrum (except for the too high intensity of the line at 414.55 Hz in the X part, which is probably due to some impurity).

The other possible parameter set is calculated according to the prescription for set (*ii*) in Section IV.4.f above, *i.e.* with $\Delta_1 = -2.99$ Hz and $\Delta_2 = 4.03$ Hz. We find (in Hz) $J_{AB} = 3.26$, $\nu_A = 352.67$, $\nu_B = 353.15$, $J_{AX} = -2.23$ and $J_{BX} = 5.09$. The X part of the spectrum would with this parameter set consist of one line pair of unit intensity separated by a frequency interval of 2.86 and one line pair of relative intensity 0.47 separated by a frequency interval of 0.76 Hz. The third line pair of separation 9.60 should in this scheme have a relative intensity of 0.53.

A comparison with the experimental part of the spectrum shows that the latter parameter set can be ruled out. Detailed studies have shown that in general $J_{34} > J_{45} > J_{35}$ in 2-substituted furans, and we can therefore assign $H_A = H_3$, $H_B = H_4$ and $H_X = H_5$.

Example 2: The Proton and Fluorine Resonance Spectra of 2-fluoro-4-chlorothiophene

The proton and fluorine spectra of the title compound obtained at a fixed frequency of 56.4 MHz is depicted in Fig. 34 (note the change in the H_0 field between the two spectra). The proton spectrum shows overlap and degeneracies of lines, only five of the conceivable eight lines are clearly visible. The three low-field lines may be assigned to an *ab* subspectrum where the fourth line is hidden in the strongest line in the spectrum. With this assumption we find $a_1 = 4.74$ Hz and $b_1 = 0.63$ Hz. It is then apparent that the other *ab* subspectrum is very strongly coupled, and its two central lines almost coincide at the frequency 453.46 Hz from TMS. Only one of the outer lines in this *ab* subspectrum can be seen at the frequency 451.34 Hz from TMS.

Since we cannot resolve the two central lines in this *ab* subspectrum we put $b_2 = 0.00$ Hz, while $a_2 = 4.26$ Hz. The centre frequencies of the two *ab* subspectra are taken as $\bar{\nu}_1 = 455.64$ Hz and $\bar{\nu}_2 = 453.46$ Hz from TMS. Proceeding with the step-by-step analysis, we calculate $\Delta_1 = 1.73$ Hz and $\Delta_2 = 0.00$ Hz. It is now realized that, since $\Delta_2 = 0.00$ Hz, parameter set (*i*) equals that of set (*ii*). The parameters calculated from Eqs. (26)–(30) are given by (in Hz) $J_{AB} = 2.09$ (average value), $\nu_A = 454.98$, $\nu_B = 454.12$, $J_{AX} = 3.05$ and $J_{BX} = 1.31$. Before calculating the X part of the spectrum we have to correct for the different H_0-fields employed in recording the proton and fluorine spectra. This may be done by means of Eqs. (31)–(32).

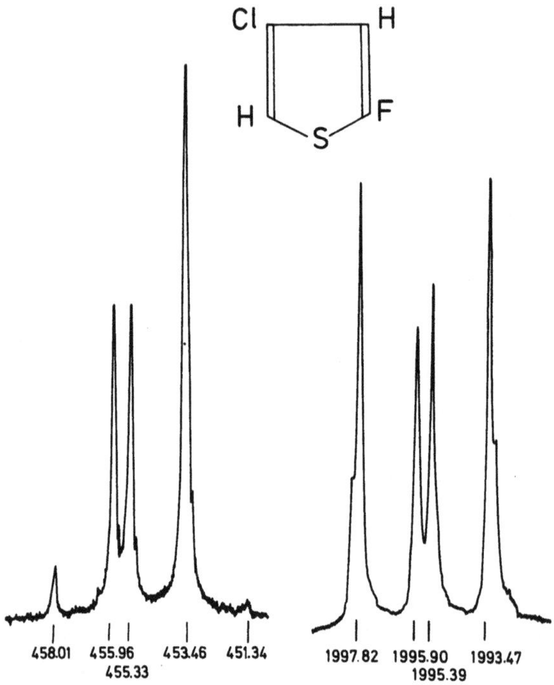

Fig. 34. The proton and fluorine magnetic resonance spectra at 56.4 MHz of 2-fluoro-4-chlorothiophene (25% by weight in cyclohexane). The line frequencies are given relative to TMS as internal reference for the proton band, and relative to hexafluorobenzene as internal reference for the fluorine band

Since a fixed frequency has been used, the ratio of the magnetic fields H'_0/H_0 simply equals $\gamma_H/\gamma_F = 1.063$. This gives $a'_1 + b'_1 = 5.49$ Hz and $a'_2 + b'_2 = 4.18$ Hz. This would give an X spectrum consisting of one line pair of unit intensity at a separation of $|J_{AX} + J_{BX}| = 2|\bar{\nu}_1 - \bar{\nu}_2| = 4.36$ Hz, a second line pair with a splitting $|a'_1 + b'_1 - a'_2 - b'_2|/2 = 0.66$ Hz and a third line pair with a splitting $|a'_1 + b'_1 + a'_2 + b'_2|/2 = 4.79$ Hz. The relative intensities of the latter doublets are, in accordance with step 6, found to be 0.88 and 0.12 respectively.

A comparison of the calculated and experimental line splittings shows fair agreement; however, the experimental splitting in the inner line pair is 0.15 Hz smaller than the calculated splitting, while the experimental splitting in the outer line pair is larger than the calculated one. This discrepancy indicates that our original assumption of $b_2 = 0$ is not entirely correct. In fact, under very high resolution, the central doublet in the high field *ab* subspectrum can be resolved, showing a splitting of 0.11 Hz. Small though this value seems, it has the effect of changing the calculated coupling constants into $J_{AB} = 2.07$ Hz, $J_{AX} = 3.40$ Hz and $J_{BX} = 0.95$ Hz. This amply illustrates the difficulty of obtaining an accurate value for the chemical shift in a strongly coupled AB spectrum, as discussed in Section IV.2c. From the spectrum alone an assignment of the A and B proton cannot be made. From studies of other fluorinated thiophenes it has been established, however, that $J_{F-5} > J_{F-3}$ (S. Rodmar, to be published). We can therefore conclude that $\nu_A = \nu_5$ and $\nu_B = \nu_3$.

IV.5. Generalization of the ABX Analysis Method: ABPX, ABX_n etc.

a) The Method of Effective Larmor Frequencies

The AB basic multiplet in spin systems of the type ABPX, ABX_n, $ABPX_n$ etc. (*i.e.* spin systems with one single pair of nonequivalent strongly coupled nuclei) may, just as in the ABX case, be divided into a number of AB-type subspectra, the number of subspectra being given by the number of combinations of magnetic quantum numbers of the other basic groups. Thus the AB basic multiplet of an ABPX spectrum will consists of $2 \times 2 = 4ab$ subspectra, that of an ABX_2 spin system will consist of $3ab$ subspectra and that of an $ABPX_3$ spin system will consist of $2 \times 4 = 8ab$ subspectra (cf. Table 2).

The total intensity of the AB basic multiplet will be distributed over the *ab* subspectra according to the statistical weights of the corresponding magnetic quantum numbers. Thus, the three *ab* subspectra of an ABX_2 spin system will have the relative intensities 1 : 2 : 1, these intensities being obtained as the sums of the four lines in each *ab* subspectrum. The apparent spin-coupling constant, J_{ab}, is the same for all subspectra and equals the true spin-coupling constant J_{AB}, while the apparent chemical shifts are given by the effective Larmor frequencies. Thus, the apparent chemical shifts of the *ab* subspectra in an $ABPX_3$ spin system are given by the formulas

$$\nu_a = \nu_A + J_{AP} M_P + J_{AX} M_X \tag{42}$$

$$\nu_b = \nu_B + J_{BP} M_P + J_{BX} M_X \tag{43}$$

where M_P can assume the two values $M_P = +1/2$ and $M_P = -1/2$, while M_X can assume the four values of $M_X = +3/2$, $M_X = +1/2$, $M_X = -1/2$ and $M_X = -3/2$, respectively.

b) Basic Structure of an ABX_3 Spectrum

The basic structure of ABX_n spectra will be discussed with reference to an ABX_3 spin system for the sake of concreteness. A complication which arises from the existence of several magnetically equivalent nuclei in the X-groups is that the spin state of that group is no longer adequately represented by its magnetic quantum number, M_X, unless $|M_X|$ assumes its maximum value of 3/2. This is so because each of the magnetic quantum numbers $M_X = \pm 1/2$ corresponds to three possible states among the X nuclei (cf. Table 2). The effect of this degeneracy on the AB basic multiplet is simply taken into account by assigning to each *ab* subspectrum belonging to $M_X = +1/2$ of $-1/2$ a statistical weight of three (cf. Table 2). This simple device cannot be applied for evaluating the intensities of the lines in the X basic multiplet, however, but fortunately these latter relative intensities are not really required for an unambiguous assignment of chemical shifts and spin-coupling constants. A full treatment of the intensity distribution in the X basic multiplet can be given in terms of the composite particle model to be discussed in Section IV.6 but will not be elaborated here, and our discussion will be confined to those features of ABX_n spectra which are required to derive the NMR parameters. What can be obtained with this treatment is the magnitudes of J_{AB}, J_{AX}, J_{BX} and the relative signs of the two latter couplings in addition to ν_A, ν_B and ν_X.[8]

The minor simplification that we shall employ is to characterize the states of the X spins merely by their magnetic quantum number M_X, and those levels of the spin system that become *exactly* degenerate in spin energy when the unobservable mutual couplings among nuclei within the X_n group are set equal to zero will be drawn as a single energy level in our diagrams. The ABX_3 spin system then obtains altogether 16 different energy levels, since M_X can assume the four values of $+3/2$, $+1/2$, $-1/2$ and $-3/2$ while M_{AB} can assume the values of $+1$, -1 and 0, the value $M_{AB}=0$ being obtainable in two different ways so that altogether $4\times 4=16$ possible combinations of the magnetic quantum numbers are possible.

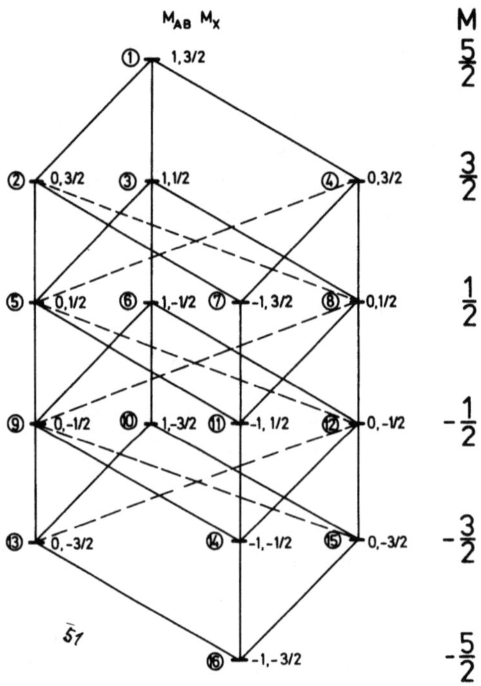

Fig. 35. Energy-level and transition diagram of an ABX_3 spin system. The pair of numbers next to each level gives the magnetic quantum numbers M_{AB} and M_X for each state respectively. The broken lines show the transitions which become forbidden in the APX_3 limit

A schematic energy level and transition diagram for an ABX_3 spin system is depicted in Fig. 35. In this diagram the four *ab* subspectra are represented by the following four quadrupoles, $M_X=3/2$: 1-2-7-4; $M_X=1/2$: 3-5-11-8; $M_X=-1/2$: 6-9-14-12; $M_X=-3/2$: 10-13-16-15. The total AB basic multiplet intensity will

[8] No additional NMR parameters would be obtained from a more rigorous treatment, but – naturally – the redundant information could be utilized as a check and for increasing the accuracy.

be shared among these subspectra in the ratio 1:3:3:1, and the relative intensities of the lines in any one *ab* subspectrum are those of a genuine AB spectrum. The apparent spin couplings will all be the same and equal to J_{AB}. Thus, once the subspectra have been identified, the value of J_{AB} is immediately obtained.

The effective Larmor frequencies in the four subspectra will be given by

$$\begin{aligned} \nu_a &= \nu_A + J_{AX} M_X \\ \nu_b &= \nu_B + J_{BX} M_X \end{aligned} \tag{44}$$

with $M_X = +3/2, \; +1/2, \; -1/2, \; -3/2$, and these four subspectra will be labelled by numbers 1 to 4 in the order of decreasing M_X. The midpoint of subspectrum number i will be written $\bar{\nu}_i$ and the apparent chemical shift will be written Δ_i. They are defined as

$$\begin{aligned} \bar{\nu} &= (\nu_a + \nu_b)/2 \\ \Delta &= \nu_a - \nu_b \,. \end{aligned} \tag{45}$$

By the use of Eqs. (44) and (45) with $i = 1$ to 4, the following relationships are easily derived

$$\bar{\nu}_1 = (\nu_A + \nu_B)/2 + 3(J_{AX} + J_{BX})/4\,, \tag{46a}$$

$$\bar{\nu}_2 = (\nu_A + \nu_B)/2 + (J_{AX} + J_{BX})/4\,, \tag{46b}$$

$$\bar{\nu}_3 = (\nu_A + \nu_B)/2 - (J_{AX} + J_{BX})/4\,, \tag{46c}$$

$$\bar{\nu}_4 = (\nu_A + \nu_B)/2 - 3(J_{AX} + J_{BX})/4\,, \tag{46d}$$

$$\Delta_1 = \nu_A - \nu_B + 3(J_{AX} - J_{BX})/2\,, \tag{46e}$$

$$\Delta_2 = \nu_A - \nu_B + (J_{AX} - J_{BX})/2\,, \tag{46f}$$

$$\Delta_3 = \nu_A - \nu_B - (J_{AX} - J_{BX})/2\,, \tag{46g}$$

$$\Delta_4 = \nu_A - \nu_B - 3(J_{AX} - J_{BX})/2\,. \tag{46h}$$

From Eqs. (46a) to (46d) it follows that the centers of the *ab* subspectra will be displaced evenly by the amount

$$\bar{\nu}_1 - \bar{\nu}_2 = \bar{\nu}_2 - \bar{\nu}_3 = \bar{\nu}_3 - \bar{\nu}_4 = (J_{AX} + J_{BX})/2 \tag{47a}$$

and that the apparent chemical shift values will obey a sum rule

$$\Delta_1 + \Delta_4 = \Delta_2 + \Delta_3\,, \tag{47b}$$

$$\Delta_1 - \Delta_4 = 3(\Delta_2 - \Delta_3)\,. \tag{47c}$$

Since the signs of J_{AX} and J_{BX} are determined only relative to one another, one may arbitrarily assign a positive sign to the larger one of J_{AX} and J_{BX} so that all numbers of Eq. (47a) become positive:

$$\bar{\nu}_1 > \bar{\nu}_2 > \bar{\nu}_3 > \bar{\nu}_4\,. \tag{48}$$

Furthermore, since the labels A and B are arbitrary, one may without lack of generality assume that $\nu_A > \nu_B$ and it then follows from Eqs. (46e) to (46h) that

$$\Delta_1 + \Delta_2 + \Delta_3 + \Delta_4 > 0. \tag{49}$$

When the four *ab* subspectra have been identified in the AB basic multiplet, Eq. (48) allows one to assign a definite number to every subspectrum, *i.e.* the assignment of a definite value of M_X to every observed subspectrum. The Δ_i values are obtained by use of the analysis procedure described in Section IV.2.b to the *ab* subspectra and the signs of the Δ_i:s are made to conform with Eqs. (47b) and (49). It will be noted that the sign of at least one of the Δ_i:s will be negative if $3\,|J_{AX} - J_{BX}| > 2\,|\nu_A - \nu_B|$ (cf. Eqs. (46e) to (46h)). The four unknowns ν_A, ν_B, J_{AX} and J_{BX} on the right-hand sides of Eqs. (46a) to (46h) may then be obtained from the measured values of $\bar{\nu}_i$ and Δ_i by solving the equations. Since the number of equations vastly exceeds the number of unknowns, the analytical expressions obtained depend on the method of solution. The following expressions are obtained by giving one and the same weight to the information of any one subspectrum. Obviously, if any of the four subspectra are of poor quality, alternative expressions for the NMR parameters ought to be derived from Eqs. (46a) to (46h).

$$\nu_A = (\bar{\nu}_1 + \bar{\nu}_2 + \bar{\nu}_3 + \bar{\nu}_4)/4 + (\Delta_1 + \Delta_2 + \Delta_3 + \Delta_4)/8\,, \tag{50a}$$

$$\nu_B = (\bar{\nu}_1 + \bar{\nu}_2 + \bar{\nu}_3 + \bar{\nu}_4)/4 - (\Delta_1 + \Delta_2 + \Delta_3 + \Delta_4)/8\,, \tag{50b}$$

$$J_{AX} = (\bar{\nu}_1 + \bar{\nu}_2 - \bar{\nu}_3 - \bar{\nu}_4)/4 + (\Delta_1 + \Delta_2 - \Delta_3 - \Delta_4)/8\,, \tag{50c}$$

$$J_{BX} = (\bar{\nu}_1 + \bar{\nu}_2 - \bar{\nu}_3 - \bar{\nu}_4)/4 - (\Delta_1 + \Delta_2 - \Delta_3 - \Delta_4)/8\,. \tag{50d}$$

Thus, by analysis of the four *ab* subspectra one obtains the values of J_{AB}, ν_A, ν_B, J_{AX}, J_{BX} and the relative signs of J_{AX} and J_{BX}. With the conventions of Eqs. (48) and (49) the analysis will always yield $\nu_A > \nu_B$ and a positive sign to the larger one of J_{AX} and J_{BX}. It will be noted that all NMR parameters, except ν_X, can be obtained from an analysis of the AB basic multiplet alone. This is true in general in ABX_n spectra with $n > 1$, in contrast to the ABX case where the ambiguity as to which one is which of $\nu_A - \nu_B$ and $(J_{AX} - J_{BX})/2$ could be resolved solely by analysis of the X basic multiplet.

It may be shown that the X basic multiplet will always display mirror-image symmetry and be centered at the Larmor frequency ν_X of X. The most prominent feature in the X basic multiplet is a pair of lines separated by a splitting of $|J_{AX} + J_{BX}|$ and each one of these lines has an intensity of 1/4 of the total X basic multiplet intensity. There will be no lines in the X basic multiplet that are stronger than these two lines. In the diagram of Fig. 35 these two strong lines correspond to transitions between unmixed states and comprise the degenerate transitions 1–3, 3–6, 6–10 (all having the frequency $\nu_X + (J_{AX} + J_{BX})/2$) and 7–11, 11–14, 14–16 (with frequency $\nu_X + (J_{AX} + J_{BX})/2$), respectively. There will also be three pairs of lines, which in a weakly coupled first order spectrum would degenerate into one single pair split by $|J_{AX} - J_{BX}|$. In the diagram of Fig. 35 these lines are represented by the transitions 2–5, 5–9, 9–13 and 4–8, 8–12, 12–15 respectively. These lines all correspond to transitions between mixed states. Their transition frequencies may be obtained from the level diagram by use of frequency sum rules along any closed quadrupole. Finally, there exist three pairs of so-called combination transitions, indicated in Fig. 35 by dashed lines. They correspond to transitions between levels 2–8, 5–12, 9–15, 4–5, 8–9 and 12–13. They also involve transitions between mixed states and their transition frequencies may also be derived by frequency sum rules. In terms of the frequency intervals measured in the AB basic multiplet, the splittings between pairs of transitions involving mixed states are given by generalization of Eqs. (35) and (36) of Section IV.4.f. If the quality of the AB basic multiplet is good enough to yield unambiguous values for all the parameters obtainable, the only additional information one can gain by analysing the X basic multiplet is the value of ν_X. In certain cases, however, the relative signs of J_{AX} and J_{BX} are more readily obtainable from the appearance of the X basic multiplet.

c) Basic Structure of an ABX_2 Spectrum

A schematic energy-level and transition diagram is depicted in Fig. 36. In drawing this figure, we employed the same simplification as used in the diagram of the ABX_3 spin system of Fig. 35, by superimposing degenerate energy levels. Thus all energy levels in which $M_X = 0$ are doubly degenerate, *viz.* levels number 3, 5, 8, and 10 in Fig. 36. Most of the following discussion will be based on the conventions followed in analysing the ABX_3 spin system, and the notation of subsection (b) is taken over with obvious modifications. The AB basic multiplet decomposes into three *ab* subspectra with the relative intensities 1:2:1, and an analysis of these subspectra yields the values of J_{AB}, $\nu_A - \nu_B$, J_{AX}, J_{BX} and the relative signs of the latter two coupling constants. By assigning to the three

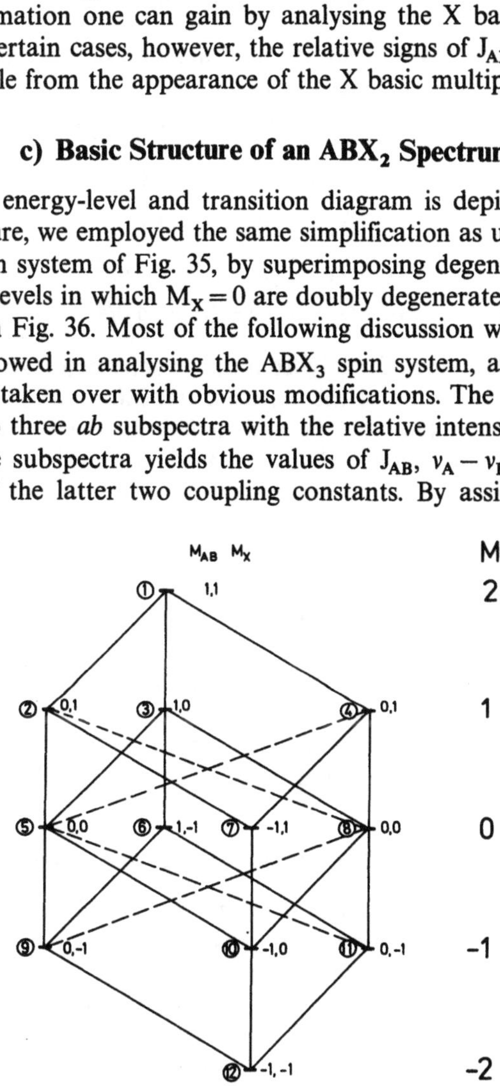

Fig. 36. Energy-level and transition diagram of an ABX_2 spin system. The pair of numbers next to each level gives the magnetic quantum numbers M_{AB} and M_X for each state respectively. The broken lines show the transitions which become forbidden in the APX_2 limit

subspectra labels number 1 to 3 with the convention that

$$\bar{\nu}_1 > \bar{\nu}_2 > \bar{\nu}_3 \tag{51}$$

and

$$\Delta_1 + \Delta_2 + \Delta_3 > 0 \tag{52}$$

one will obtain NMR parameters such that $\nu_A > \nu_B$ and $J_{AX} + J_{BX} > 0$. The parameters $\bar{\nu}_i$ and Δ_i obey the following equations

$$\bar{\nu}_1 - \bar{\nu}_2 = \bar{\nu}_2 - \bar{\nu}_3 > 0\,, \tag{53}$$

$$2\Delta_2 = \Delta_1 + \Delta_3 > 0\,. \tag{54}$$

The *ab* subspectra are easily identified, and analysis of these by the use of the procedures given in Section IV.2.b gives the values of the ν_i:s and Δ_i:s. The NMR parameters are given in terms of these experimentally obtained entities by the expressions

$$\bar{\nu}_A = (\bar{\nu}_1 + \bar{\nu}_2 + \bar{\nu}_3)/3 + (\Delta_1 + \Delta_2 + \Delta_3)/6\,, \tag{55a}$$

$$\nu_B = (\bar{\nu}_1 + \bar{\nu}_2 + \bar{\nu}_3)/3 - (\Delta_1 + \Delta_2 + \Delta_3)/6\,, \tag{55b}$$

$$J_{AX} = (\bar{\nu}_1 - \bar{\nu}_2)/2 + (\Delta_1 - \Delta_3)/4\,, \tag{55c}$$

$$J_{BX} = (\bar{\nu}_1 - \bar{\nu}_3)/2 - (\Delta_1 - \Delta_3)/4\,. \tag{55d}$$

and J_{AB} is, as usual, obtained directly from the *ab* subspectra. As was the case for the ABX_3 spin system, some of the experimental information is redundant, as may be seen from Eqs. (54) to (55), and in fact all the NMR parameters of Eqs. (55) may be obtained from only two of the three *ab* subspectra.

The X basic multiplet will be symmetrical with respect to ν_X. The two strongest lines in the spectrum which together make up 1/2 of the X basic multiplet intensity will be split by $|J_{AX} + J_{BX}|$. In the energy-level diagram they correspond to the pairwise degenerate transitions 1–3 and 3–6 (with resonance frequency of $\nu_X + (J_{AX} + J_{BX})/2$) and the pair 7–10 and 10–12 (with a resonance frequency of $\nu_X - (J_{AX} + J_{BX})/2$). The remaining X basic multiplet intensity is shared among four pairs of symmetrical doublets, none of which will be degenerate in the most general case. The transition frequencies of these lines may be obtained from Fig. 36 by the use of frequency sum rules. The only new NMR parameter that may be obtained from the X basic multiplet is the value of ν_X.

d) Basic Structure of an ABPX Spin System

The energy level and transition diagram of an ABPX spin system (with exclusion of the four combination transitions in the P basic multiplet) is displayed in Fig. 37. From the level diagram it is evident that the basic AB multiplet will consist of four *ab* subspectra, one for each combination of the possible values of $M_P = \pm 1/2$ and $M_X = \pm 1/2$. Alternatively, it may be advantageous to consider the AB *and* X basic multiplets together as two superimposed *abx* subspectra, one originating in molecules having $M_P = +1/2$ and the other originating in molecules having $M_P = -1/2$. In the ABX analysis of the two *abx* subspectra one would then employ the methods described in Section IV.4, but the Larmor frequencies derived would be the effective Larmor frequencies $\nu_i \pm J_{iP}/2$ (i = A, B, X) instead of simply ν_i. Similarly, the AB and P basic multiplets might be devided into a different pair of *abp* subspectra, one for each value of M_X. In practice it turns out, however, that one may derive more information by taking into consideration the relationship that must exist between the four *ab* subspectra in the AB basic multiplet. This will now be borne out.

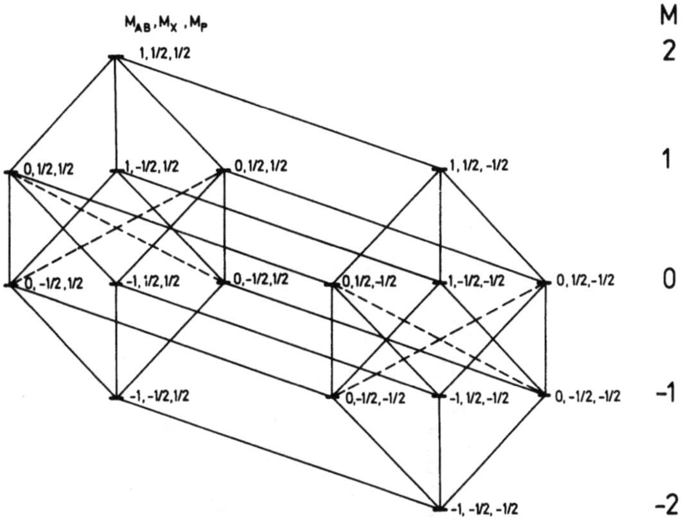

Fig. 37. Energy-level and transition diagram of an ABPX spin system. The level diagram has been drawn in such a way that the two polygons corresponding to the two *abx* subspectra are clearly displayed. The P transitions are given by the long sloping lines which connect these two *abx* subdiagrams. Only the "combination" transitions in the X part of the spectrum are shown as broken lines. The set of numbers next to each energy level shows the magnetic quantum numbers M_{AB}, M_X and M_P in each state

The midpoints $(\nu_a + \nu_b)/2$ of the four subspectra will be called $\bar{\nu}_1$ to $\bar{\nu}_4$ and the corresponding effective chemical shifts, $\nu_a - \nu_b$, will be labelled Δ_1 to Δ_4, and are listed in Table 3. From Table 3 it is evident that the centre frequencies, $\bar{\nu}_i$, obey a sum rule, *viz.*

$$\bar{\nu}_1 + \bar{\nu}_4 = \bar{\nu}_2 + \bar{\nu}_3 \tag{56}$$

and the apparent chemical shifts, Δ_i, are related by the equation

$$\Delta_1 + \Delta_4 = \Delta_2 + \Delta_3 \,. \tag{57}$$

By use of Table 3, the NMR parameters may be expressed in terms of the quantities $\bar{\nu}_i$ and Δ_i and one may verify that the following equations obtain.

$$\nu_A = (\bar{\nu}_1 + \bar{\nu}_2 + \bar{\nu}_3 + \bar{\nu}_4)/4 + (\Delta_1 + \Delta_2 + \Delta_3 + \Delta_4)/8 \,, \tag{58a}$$

$$\nu_B = (\bar{\nu}_1 + \bar{\nu}_2 + \bar{\nu}_3 + \bar{\nu}_4)/4 - (\Delta_1 + \Delta_2 + \Delta_3 + \Delta_4)/8 \,, \tag{58b}$$

$$J_{AP} = (\bar{\nu}_1 + \bar{\nu}_3 - \bar{\nu}_2 - \bar{\nu}_4)/2 + (\Delta_1 + \Delta_3 - \Delta_2 - \Delta_4)/4 \,, \tag{58c}$$

$$J_{BP} = (\bar{\nu}_1 + \bar{\nu}_3 - \bar{\nu}_2 - \bar{\nu}_4)/2 - (\Delta_1 + \Delta_3 - \Delta_2 - \Delta_4)/4 \,, \tag{58d}$$

$$J_{AX} = (\bar{\nu}_1 + \bar{\nu}_2 - \bar{\nu}_3 - \bar{\nu}_4)/2 + (\Delta_1 + \Delta_2 - \Delta_3 - \Delta_4)/4 \,, \tag{58e}$$

$$J_{BX} = (\bar{\nu}_1 + \bar{\nu}_2 - \bar{\nu}_3 - \bar{\nu}_4)/2 - (\Delta_1 + \Delta_2 - \Delta_3 - \Delta_4)/4 \,. \tag{58f}$$

Table 3. *Center frequencies and apparent chemical shifts of ab subspectra in an ABPX spin system*

i \ Center frequency, $\bar{\nu}_i$[a]	Apparent chemical shift, Δ_i[b]	M_X	M_P
$\nu_A/2+\nu_B/2+J_{AX}/4+J_{BX}/4+J_{AP}/4+J_{BP}/4$	$\nu_A-\nu_B+J_{AX}/2-J_{BX}/2+J_{AP}/2-J_{BP}/2$	+1/2	+1/2
$\nu_A/2+\nu_B/2+J_{AX}/4+J_{BX}/4-J_{AP}/4-J_{BP}/4$	$\nu_A-\nu_B+J_{AX}/2-J_{BX}/2-J_{AP}/2+J_{BP}/2$	+1/2	−1/2
$\nu_A/2+\nu_B/2-J_{AX}/4-J_{BX}/4+J_{AP}/4+J_{BP}/4$	$\nu_A-\nu_B-J_{AX}/2+J_{BX}/2+J_{AP}/2-J_{BP}/2$	−1/2	+1/2
$\nu_A/2+\nu_B/2-J_{AX}/4-J_{BX}/4-J_{AP}/4-J_{BP}/4$	$\nu_A-\nu_B-J_{AX}/2+J_{BX}/2-J_{AP}/2+J_{BP}/2$	−1/2	−1/2

[a] The center frequency, $\bar{\nu}_i$, is defined by $\bar{\nu}=\nu_a/2+\nu_b/2$ with ν_a and ν_b given by Eqs. (42) and (43) and values of M_P and M_X, as listed in the last two columns above.

[b] The apparent chemical shift, Δ_i, is defined by $\Delta=\nu_a-\nu_b$ with ν_a and ν_b given as in (a) above.

This is not the only way of writing the solutions since the six unknowns on the left-hand sides of Eq. (58a) through (58f) may be expressed in more than one way in terms of the eight parameters $\bar{\nu}_1$ through $\bar{\nu}_4$ and Δ_1 through Δ_4. The parameters are thus overdetermined, as is obvious also from the existence of the sum rules of Eqs. (56) and (57). This redundancy allows one to obtain the value of $\nu_A - \nu_B$ from an analysis of the AB basic multiplet alone without consideration of the P and X basic multiplets.

The appearance of an ABPX spectrum is always independent of

(i) the sign of $\nu_A - \nu_B$ (as is self-evident, since this sign depends only on our personal preference in chosing labels A and B);

(ii) the sign of $|J_{AX} + J_{BX}| - |J_{AP} + J_{BP}|$ (as is again self-evident, since this sign may be reversed by interchanging the labels P and X);

(iii) the sign of J_{AB} (as is not self-evident, but nevertheless true for all spectra of the kind enumerated in Section IV.5.a above);

(iv) the sign of one additional spin-coupling parameter (as follows from the general statement that absolute signs of spin couplings remain undetermined by the analysis).

In the analysis procedure to be described, these sign ambiguities are lifted by *arbitrarily assigning a positive sign* (i) to $\nu_A - \nu_B$; (ii) to $|J_{AX} + J_{BX}| - |J_{AP} + J_{BP}|$; (iii) to J_{AB}; and (iv) to $J_{AX} + J_{BX}$ (*i.e.* to the larger one of J_{AX} and J_{BX}). By use of these sign conventions, it becomes possible to reduce the number of possible assignments of labels 1 through 4 in the experimentally observed *ab* subspectra to *two* different sets corresponding either to

$$\bar{\nu}_1 > \bar{\nu}_2 > \bar{\nu}_3 > \bar{\nu}_4 \tag{59}$$

or to

$$\bar{\nu}_2 > \bar{\nu}_1 > \bar{\nu}_4 > \bar{\nu}_3 \,. \tag{60}$$

For a proof of this statement, the reader is referred to Appendix A.

Finally, the condition that $\nu_A - \nu_B$ should be positive may be written

$$\Delta_1 + \Delta_2 + \Delta_3 + \Delta_4 > 0 \tag{61}$$

since the left-hand side of (61) is readily shown from Eqs. (58e) and (58f) to equal $4(\nu_A - \nu_B)$, which entity was taken as positive by the convention (i) above. The sum rule of Eq. (57) and the inequality (61) suffice to determine the signs of all Δ_i values in a genuine ABPX spectrum.

e) Analysis Procedure for an ABPX Spectrum

The first step in a practical analysis of an experimental ABPX spectrum is to decompose the AB basic multiplet into four *ab* subspectra. Unless the AB basic multiplet shows degeneracies owing to accidental overlap or deceptive simplicity, this step may be performed on inspection by noting that the intensity sum of the four lines in any one *ab* subspectrum must equal 1/4 of the total AB basic multiplet intensity and that every *ab* subspectrum has the appearance of a genuine AB spectrum with a spin-coupling constant J_{AB}, so that this splitting should recur eight times in the AB basic multiplet. The center frequency, $\bar{\nu}_i$, and the apparent chemical shift Δ_i are then evaluated for every subspectrum, and the labelling of subspectra $i=1$ to 4 is made in accordance with Eq. (59) or (60) and the signs of the Δ_i:s are made to conform to Eqs. (57) and (61). One thus obtains two sets of parameters, *set I* being obtained by labelling according to Eq. (59) and *set II* by labelling according to Eq. (60). In set I the sign of $J_{AP}+J_{BP}$ is the same as that of $J_{AX}+J_{BX}$, and hence positive according to our convention, whereas in set II the signs of J_{AP} and J_{BP} are opposite to those of set I. Unless one has reason to suppose that the sign of $J_{AB}+J_{BP}$ is opposite to that of $J_{AX}+J_{BX}$, it is always advisable to start the analysis with the assumption that set I is the correct one, since set II only implies that some of the signs of some spin couplings have to be reversed but does not entail any change in the magnitude of the parameters, in contrast to the situation in the ABX spin system.

The final decision as to which of the two sets is the correct one can only be made in conjunction with an analysis of both the P and the X basic multiplets, and even then only if J_{PX} produces observable splittings. Otherwise the analysis will only yield the sign of J_{AX} relative to J_{BX} and the sign of J_{AP} relative to that of J_{BP}, but the relative signs of the two pairs will remain undetermined. According to our conventions, the X basic multiplet must be assigned to that weakly coupled group which displays the larger coupling sum, so that $|J_{AX}+J_{BX}|>|J_{AP}+J_{BP}|$. This basic multiplet will be a superposition of *two* symmetrical six-line patterns, each one of which has the same structure as that of an X basic multiplet in a simple ABX three-spin system. Since the nucleus P is weakly coupled to the rest of the spin system, the effective Larmor frequency approach again commends itself. By disregarding for the moment the P basic multiplet, we may regard the spectrum as a superposition of two ABX spectra (or *abx* subspectra) in which the effective Larmor frequencies are given by

$$\nu_a=\nu_A+J_{AP}M_P\,, \tag{62}$$

$$\nu_b=\nu_B+J_{BP}M_P\,, \tag{63}$$

$$\nu_x=\nu_X+J_{PX}M_P\,. \tag{64}$$

With $M_P=+1/2$ one obtains one *abx* subspectrum, and with $M_P=-1/2$ one obtains the other *abx* subspectrum. The two symmetrical six-line patterns in the X basic multiplet correspond to the X part subspectrum of these two sets of subspectra, and according to Eq. (64) the centers of these two X subspectra will be displaced by J_{PX}. Eqs. (62) to (64) show that the magnitude of ν_X is related to the magnitude of $\nu_a-\nu_b$ for, if $M_P=+1/2$, one obtains $x_x=\nu_X+J_{PX}/2$ and $\nu_a-\nu_b=\nu_A-\nu_B+(J_{AP}-J_{BP})/2$ and conversely, if $M_P=-1/2$, one obtains $\nu_x=\nu_X-J_{PX}/2$ and $\nu_a-\nu_b=\nu_A-\nu_B-(J_{AP}-J_{BP})/2$. This linking enables one to determine the sign of J_{PX} relative to the sign of $J_{AP}-J_{BP}$. Since the signs of J_{AP} and J_{BP} within set I (and the opposite signs in set II) are obtained by analysis of the AB basic multiplet, it follows that analysis of the X basic multiplet will yield one sign of J_{PX} within set I and the opposite sign in set II. By similar arguments, it follows that an analysis of the P basic multiplet will yield the sign of J_{PX} relative to the signs of J_{AX} and J_{BX}, and, since the sign of $J_{AX}+J_{BX}$ was by convention chosen to be positive, it follows that analysis of the X basic multiplet will enable one to decide which one of the two sets (I or II) agrees with the entire ABPX spectrum.

Example 1: The Proton and Fluorine Resonance Spectrum of 5-fluoro-3-thiolene-2-one (1)

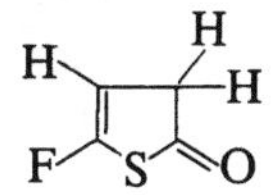

(I)

The proton resonance spectrum of (I) at 56.4 MHz is displayed in Fig. 38 and consists of two distinct bands with equal integrated intensity. Since the compound is known to contain three protons and one fluorine, it appears that each 12-line pattern must be assigned to $1\frac{1}{2}$ protons, and thus one proton has its intensity divided equally among the two bands. (It will be noted that, if nothing were known about the composition of the compound, the two proton bands might have been interpreted as two basic multiplets assigned to two basic groups.)

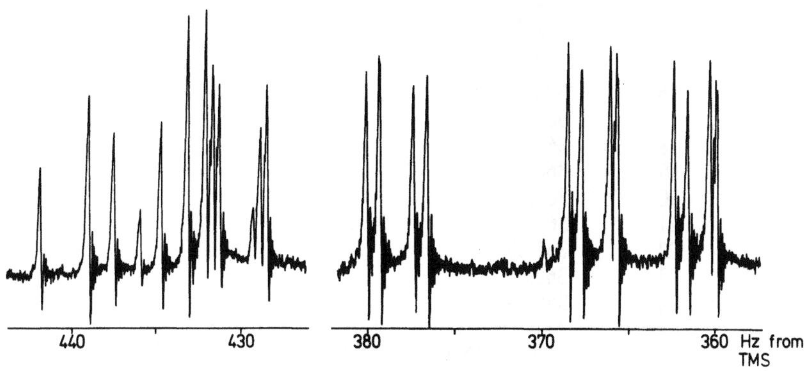

Fig. 38. The proton magnetic resonance spectrum of 5-fluoro-3-thiolene-2-one in an acetone and deuteropyridine solution

The fluorine resonance spectrum at 56.4 MHz consists of eight lines of equal intensity which may be grouped into two quartets with the separation between quartets of ca. 50 Hz and a width of each quartet of the order of ca. 6.5 Hz. The large splitting of 50 Hz is of the order expected for a geminal H–F spin coupling. The proton and fluorine transition frequencies are listed in Table 4.

The intensity distribution in the low-field proton band reveals that the spin system is not weakly coupled and hence the spin system must be classified either as an ABCX- or as an ABPX-type system. To a very good approximation, the spectrum may indeed be classified as an ABPX spin system.

The first steps in an analysis procedure will then be to identify the four *ab* subspectra and, in the present case, to identify the AB and P basic multiplets which have become somewhat intermingled owing to the large geminal H–F spin coupling. From the intensity distribution and repeated spacings among lines 4, 6, 7, and 10, one may conclude that these lines form one *ab* subspectrum. The spin coupling J_{AB} thus has the value of $\nu_4 - \nu_6 = \nu_7 - \nu_{10} = 2.78 \pm 0.01$ Hz. Similarly, lines 1, 2, 8, and 11 form a second *ab* basic multiplet. The splitting of

Table 4. *Line positions in the 56.4 MHz proton and fluorine resonance spectra of 5-fluoro-3-thiolene-2-one in acetone solution*

Proton spectrum		Fluorine spectrum	
Line no.	Frequency[a] Hz	Line no.	Frequency[b] Hz
1	441.84	F1	59.51
2	439.06	F2	57.50
3	437.58	F3	55.13
4	436.04	F4	53.28
5	434.82	F5	7.38
6	433.27	F6	5.18
7	432.25	F7	3.01
8	431.80	F8	0.97
9	431.48		
10	429.46		
11	429.05		
12	428.69		
13	380.29		
14	379.54		
15	377.55		
16	376.77		
17	368.57		
18	367.80		
19	366.14		
20	365.76		
21	362.44		
22	361.68		
23	360.68		
24	359.96		

[a] Relative to TMS as internal reference.
[b] Relative to perfluorobenzene as internal reference.

$J_{AB} \simeq 2.78$ Hz recurs also between lines 3 and 5 and between lines 9 and 12 in the low-field proton spectrum. The intensity distribution among these lines, however, rules out the possibility that these four lines together form one *ab* subspectrum. Instead one must assume that lines 3 and 5 form the low field part of a "weakly coupled" *ab* subspectrum, the high field part of which must be sought among lines 13 to 24, and similarly, for lines 9 and 12, a counterpart has to be found in the high-field proton band. In addition to the two remaining *ab* "half-subspectra", the high-filed band must contain the 12 P transitions, but in fact the high-filed band consists of only 12 lines altogether. This is not surprising, however, in view of the fact that the spins are not very strongly coupled so that all combination lines are of very low intensity, and only those 8P transitions that would be allowed in a weakly coupled four-spin system are to be observed. The P basic multiplet thus must be comprised of two mutually displaced symmetrical 4-line patterns. Two kinds of such patterns are recognized in the high-field proton band: in the first, the pattern is composed of lines 13 to 18 and 21 to 22. If these lines are ascribed to the P basic multiplet, then lines 19, 20 and 23, 24 must belong to *ab* basic multiplets, but this cannot be, since the splitting of J_{AB} does not occur among the last-mentioned lines. Consequently, the P basic multiplet is comprised of lines 17 to 24 and lines 13 to 16 make up the missing halves of the *ab* subspectra.

The splitting J_{AB} occurs between lines 13 and 15 and between lines 14 and 16, and it is not obvious from the intensity distribution whether lines 13 and 15 belong to the same *ab* subspectrum as lines 3 and 5 (in which case lines 14 and 16 would be associated to the same *ab* subspectrum as lines 9 and 12), or whether the association is between lines 3, 5, 14, 16 in one *ab* subspectrum and between lines 9, 12, 13, 15 in the other *ab* subspectrum. This ambiguity can be resolved, however, by the use of Eq. (56). In the first pairing scheme (3, 5, 13, 15), one obtains the center frequencies[9] in Hz

$$\begin{array}{ll} \bar{\nu}_1 = 435.44 & \bar{\nu}_2 = 432.76 \\ \bar{\nu}_3 = 407.56 & \bar{\nu}_4 = 404.12 \end{array} \tag{65}$$

and the frequency sums will be given by

$$\begin{aligned} \bar{\nu}_1 + \bar{\nu}_4 &= 839.56 \\ \bar{\nu}_2 + \bar{\nu}_3 &= 840.32\,. \end{aligned} \tag{66}$$

In the alternative pairing scheme (3, 5, 14, 16) the center frequencies (in Hz) will be given by

$$\begin{array}{ll} \bar{\nu}_1 = 435.44 & \bar{\nu}_2 = 432.76 \\ \bar{\nu}_3 = 407.18 & \bar{\nu}_4 = 404.50 \end{array} \tag{67}$$

and the frequency sums will be given by

$$\begin{aligned} \bar{\nu}_1 + \bar{\nu}_4 &= 839.94 \\ \bar{\nu}_2 + \bar{\nu}_3 &= 839.94\,. \end{aligned} \tag{68}$$

The frequency sums of Eq. (66) are definitely outside the experimental error (which is estimated to be in the order of 0.05 Hz), and thus the center frequencies of Eq. (67) are the ones to be used.

The four effective chemical shifts Δ_1 to Δ_4 and better estimates of J_{AB} will now be obtained by analysis of the four *ab* subspectra in the manner described in Section IV.2.b. The first *ab* subspectrum consists of lines 1, 2, 8 and 11, yielding

$$\begin{aligned} a_1 &= \nu_1 - \nu_{11} &&= 12.79\ \text{Hz} \\ b_1 &= \nu_2 - \nu_8 &&= 7.26\ \text{Hz} \\ \Delta_1 &= \sqrt{a_1 b_1} &&= 9.64\ \text{Hz} \\ J_{AB} &= (a_1 - b_1)/2 &&= 2.76\ \text{Hz}\,. \end{aligned} \tag{69}$$

The second *ab* subspectrum consists of lines 4, 6, 7, and 10, yielding

$$\begin{aligned} a_2 &= \nu_4 - \nu_{10} &&= 6.58\ \text{Hz} \\ b_2 &= \nu_6 - \nu_7 &&= 1.02\ \text{Hz} \\ \Delta_2 &= \sqrt{a_2 b_2} &&= 2.59\ \text{Hz} \\ J_{AB} &= (a_2 - b_2)/2 &&= 2.78\ \text{Hz}\,. \end{aligned} \tag{70}$$

[9] The assignment of labels 1 to 4 here is made to conform with Eqs. (56) and thus corresponds to the parameter set I of an ABPX spectrum. It will be noted, however, that, in going from set I to set II, one merely interchanges labels 1 and 2 concurrently with labels 3 and 4, and this interchange will leave the sum rule of Eq. (56) invariant.

The third *ab* subspectrum consists of lines 3, 5, 14, and 16, yielding

$$\begin{aligned} a_3 &= \nu_3 - \nu_{16} &&= 60.81\text{ Hz} \\ b_3 &= \nu_5 - \nu_{14} &&= 55.28\text{ Hz} \\ \Delta_3 &= \sqrt{a_3 b_3} &&= 57.98\text{ Hz} \\ J_{AB} &= (a_3 - b_3)/2 &&= 2.76\text{ Hz}. \end{aligned} \tag{71}$$

The fourth and last *ab* subspectrum consists of lines 9, 12, 13, and 15, yielding

$$\begin{aligned} a_4 &= \nu_9 - \nu_{15} &&= 53.98\text{ Hz} \\ b_4 &= \nu_{12} - \nu_{13} &&= 48.40\text{ Hz} \\ \Delta_4 &= \sqrt{a_4 b_4} &&= 51.09\text{ Hz} \\ J_{AB} &= (a_4 - b_4)/2 &&= 2.79\text{ Hz}. \end{aligned} \tag{72}$$

By the use of Eqs. (58a) to (58f), one may evaluate the NMR parameters of set I to obtain

$$\begin{aligned} \nu_A &= 435.13\text{ Hz} \\ \nu_B &= 404.81\text{ Hz} \\ J_{AP} &= 6.17\text{ Hz} \\ J_{BP} &= -0.81\text{ Hz} \\ J_{AX} &= 4.05\text{ Hz} \\ J_{BX} &= 52.47\text{ Hz} \end{aligned} \tag{73}$$

and the mean value of J_{AB} obtained from Eqs. (69) to (72) is

$$J_{AB} = 2.77\text{ Hz}. \tag{74}$$

The parameters of set II are the same, except for the signs of J_{AP} and J_{BP}, which will both be opposite to those of Eq. (73).

We now turn to an analysis of the P basic multiplet. The two symmetrical quartets in this basic multiplet (which will henceforth be referred to as the p' and p'' subspectra of the P basic multiplet) are readily identified and are composed of lines 17, 18, 21, 22 and lines 19, 20, 23, 24, respectively[10]. From Eq. (73) the magnitude of $J_{AP} + J_{BP}$ is found to have the value of 5.36 Hz and hence the unmixed transitions in the p' subspectrum correspond to lines 18 and 21, since $\nu_{18} - \nu_{21}$ equals 5.36 Hz. The splitting $\nu_{17} - \nu_{22}$ is given by

$$\nu_{17} - \nu_{22} = 6.89\text{ Hz}. \tag{75}$$

This second splitting corresponds to the splitting between a pair of mixed transitions in the p' subspectrum and hence must be given by a formula analogous to Eq. (35) or (36) of Section IV.4.f. The transitions in the p' subspectrum correspond to transitions in molecules with a definite value of M_X (*i.e.* the

[10] In fact the p′ and p″ subspectra do not show perfect mirror-image symmetry about their midpoints, and this is indicative of a slight insufficiently in our treatment in that the spectrum departs somewhat from a genuine ABPX spectrum, tending to the ABCX type. Upon completion of the analysis, one may employ the methods described in Section IV.1.a to satisfy oneself that the second order effects are just equal to the experimental error.

magnetic quantum number of the fluorine nucleus), as do the pairs of *ab* subspectra centered at $\bar{v}_1$ and $\bar{v}_2$ *or* at $\bar{v}_3$ and $\bar{v}_4$, respectively. What one now has to do, therefore, is to compare the splitting of 6.89 Hz with the following entities

$$|a_1 + b_1 - a_2 - b_2|/2 = 6.23 \text{ Hz}, \quad (76)$$

$$|a_3 + b_3 - a_4 - b_4|/2 = 6.88 \text{ Hz}, \quad (77)$$

$$|a_1 + b_1 + a_2 + b_2|/2 = 27.65 \text{ Hz}, \quad (78)$$

$$|a_3 + b_4 + a_4 + b_4|/2 = 218.47 \text{ Hz}. \quad (79)$$

The splittings of Eq. (78) and Eq. (79) are far larger than any observed in the P basic multiplet and hence may be ascribed to unobservably weak combination transitions.

Of the two remaining entities, Eqs. (76) and (77), the value of 6.88 Hz is in excellent agreement with the value 6.89 Hz obtained in Eq. (75). Consequently, the p' subspectrum belongs to the same spin state of the fluorine as do the *ab* subspectra labelled numbers 3 and 4 (*i.e.* to $M_X = -1/2$, cf. Table 3).

In the p'' subspectrum the unmixed transitions correspond to lines 20 and 23 since

$$v_{20} - v_{23} = 5.38 \text{ Hz} \quad (80)$$

and the value of

$$v_{19} - v_{24} = 6.18 \text{ Hz} \quad (81)$$

matches within experimental error with the calculated splitting of Eq. (76). Now the effective Larmor frequencies of the p' and p'' subspectra are obtained as

$$v_{p'} = (v_{17} + v_{18} + v_{21} + v_{22})/4 = 365.12, \quad (82)$$

$$v_{p''} = (v_{19} + v_{20} + v_{23} + v_{24})/4 = 363.06, \quad (83)$$

and theoretically they are given by

$$v_{p'} = v_P - J_{PX}/2, \quad (84)$$

$$v_{p''} = v_P + J_{PX}/2 \quad (85)$$

from which one obtains

$$v_P = 364.09 \text{ Hz}, \quad (86)$$

$$J_{PX} = -2.06 \text{ Hz}. \quad (87)$$

(Note that we have departed from the analysis method described in section *d* above by analysing the P basic multiplet prior to the analysis of the X basic multiplet. The sign of J_{PX} thus has been determined relative to J_{AX} and J_{BX} and will remain negative also in parameter set II.)

The additional information obtainable from the fluorine spectrum is (i) the value of v_X, and (ii) the signs of J_{AP} and J_{BP} relative to J_{AX} and J_{BX}, since parameter sets I and II will yield different theoretical fluorine resonance spectra and hopefully only one of those will agree with the experimental one.

The fluorine resonance spectrum was recorded at the same nominal frequency as the proton resonance spectrum, so the magnetic field H_0' employed in the ^{19}F investigation differs from the field H_0 employed in obtaining the ^1H resonance spectrum, the ratio being given by

$$H_0'/H_0 = \gamma_{^1H}/\gamma_{^{19}F} = 1.063\,. \tag{88}$$

Consequently, one has to perform the calculations indicated in step (5) of Section IV.4.f. The method of calculation will be indicated for the *ab* subspectrum centered at $\bar{\nu}_1$. The pairs of *ab* subspectra which correspond to one and the same value of M_P are those centered at $\bar{\nu}_1$ and $\bar{\nu}_3$ *or* those centered at $\bar{\nu}_2$ and $\bar{\nu}_4$. From Eqs. (69) and (71) one obtains $\Delta_1 + \Delta_3 = 67.62$ Hz and $\Delta_1 - \Delta_3 = -48.34$ Hz, so that $(\gamma_H/\gamma_F)(\Delta_1 + \Delta_3 + (\Delta_1 - \Delta_3) = 23.46$ and (γ_H/γ_F) $(\Delta_1 + \Delta_3) - (\Delta_1 - \Delta_3) = 120.30$ Hz. By Eqs. (31) and (32) of Section IV.4.f, one thus obtains

$$a_1' + b_1' = 24.10\ \text{Hz}\,, \tag{89}$$

$$a_3' + b_3' = 120.42\ \text{Hz}\,. \tag{90}$$

In a similar way one obtains

$$a_2' + b_2' = 10.26\ \text{Hz}\,, \tag{91}$$

$$a_4' + b_4' = 105.62\ \text{Hz}\,. \tag{92}$$

The calculated splittings between mixed transitions in the fluorine spectrum hence are given by

$$|a_1' + b_1' - a_3' - b_3'|/2 = 48.16\ \text{Hz}\,, \tag{93}$$

$$|a_2' + b_2' - a_4' - b_4'|/2 = 47.68\ \text{Hz}\,, \tag{94}$$

$$|a_1' + b_1' + a_3' + b_3'|/2 = 72.26\ \text{Hz}\,, \tag{95}$$

$$|a_2' + b_2' + a_4' + b_4'|/2 = 57.94\ \text{Hz}\,. \tag{96}$$

In addition the splitting between unmixed transitions, viz.

$$J_{AX} + J_{BX} = 56.52\ \text{Hz} \tag{97}$$

should be retrieved twice in the fluorine spectrum. From Table 4 it is immediately evident that the only way to obtain two splittings equal to 56.52 Hz in the fluorine spectrum is given by

$$\nu_{F1} - \nu_{F7} = 56.50\ \text{Hz} \tag{98}$$

and

$$\nu_{F2} - \nu_{F8} = 56.53\ \text{Hz}\,. \tag{99}$$

The splittings of Eqs. (95) and (96) are larger than any of the remaining splittings. With a little trial and error, one then finds that the splittings which agree with those of Eqs. (93) and (94) within experimental error are those obtained below

$$\nu_{F3} - \nu_{F5} = 47.75\ \text{Hz} \tag{100}$$

and

$$\nu_{F4} - \nu_{F6} = 48.10\ \text{Hz}\,. \tag{101}$$

The two X subspectra, henceforth referred to as x' and x'', thus are comprised of lines F1, F3, F5, F7 and of lines F2, F4, F6, and F8 respectively.

Their center frequencies are given by

$$\nu_{x'} = 31.26\ \text{Hz}, \tag{102}$$

$$\nu_{x''} = 29.23\ \text{Hz}. \tag{103}$$

Since the splitting between lines F3 and F5 equals that of Eq. (94), it follows that the x' subspectrum arises from molecules in which $M_P = -1/2$ (cf. Table 3) and consequently the x'' subspectrum to $M_P = +1/2$. Hence one obtains

$$\nu_{x'} = \nu_X - J_{PX}/2, \tag{104}$$

$$\nu_{x''} = \nu_X + J_{PX}/2. \tag{105}$$

From Eqs. (102) to (105) one obtains

$$\nu_X = 30.25\ \text{Hz} \tag{106}$$

relative to the hexafluorobenzene internal reference, and

$$J_{PX} = -2.03\ \text{Hz}. \tag{107}$$

The fact that J_{PX} as obtained from the P basic multiplet (Eq. (87)) is of the same sign as that obtained from the X basic multiplet (Eq. (107)) shows that the parameter set I is indeed the correct one. Had the sign of J_{PX} in Eq. (107) been opposite to that obtained from the P basic multiplet, this would have necessitated a renumbering of the *ab* subspectra and the signs of J_{AP} and J_{BP} would then have been reversed.

The correct parameters of 5-fluoro-3-thiolene-2-one are thus those listed in Eqs. (73), (74), (86), (87), and (106). As mentioned in Section IV.5.d, the signs of J_{AB} relative to those of the other couplings, as well as the absolute signs, cannot be determined from the analysis. It may be inferred, however, from analogous compounds that the absolute signs given above are indeed the correct ones.

In conclusion it should also be pointed out that the spectra just analysed exemplify the fact that, even though the parameters

$$\lambda_{ij} = |J_{ij}/(\nu_i - \nu_j)|$$

for all i and j are small compared to unity, the spectrum need not be of first order owing to the fact that $J_{AX} - J_{BX}$ is *not* small compared to $\nu_A - \nu_B$.

IV.6. Magnetic Equivalence and Composite Particles

a) The Composite Particle Concept

It has been mentioned before that the magnetic quantum number M_A of a group of n magnetically equivalent nuclei, A_n, does not suffice to represent the spin state of the nuclei in the group. The additional quantum number required in NMR applications is the total spin quantum number F(A).

It may be shown that the total spin quantum number F(A) of a group of magnetically equivalent nuclei does not change during *any* NMR transition, so that the entire spectrum may be decomposed into subspectra according to the value of F(A). The molecules in the sample may be divided into distinct species and in any one species the n magnetically equivalent nuclei are adequately represented by a *composite particle* of total spin F(A).

The total spin quantum number F(A) is defined in the following way. Let $\boldsymbol{I}(1)$, $\boldsymbol{I}(2) \dots \boldsymbol{I}(\mathrm{n})$ represent the spin vectors of the n individual spins in the A_n group. Then we may define a resultant spin angular momentum vector, $\boldsymbol{F}(\mathrm{A})$, by the equation

$$\boldsymbol{F}(\mathrm{A}) = \boldsymbol{I}(1) + \boldsymbol{I}(2) + \dots \boldsymbol{I}(n) \qquad (108)$$

The vector $\hbar\boldsymbol{F}(\mathrm{A})$ represents a quantum mechanical angular momentum and the magnitude of $\boldsymbol{F}(\mathrm{A})$ is represented by a quantum number, F(A), which can assume only integral or half integral values. When the magnetically equivalent nuclei all have spin $\mathrm{I} = 1/2$ (and this will henceforth always be assumed), the possible values of F(A) are integral if n is even and half integral if n is odd.

Consider for example the case when $n = 2$, i.e. an A_2 grouping. When the two A nuclei have parallell spins, the resultant will have $\mathrm{F(A)} = 1/2 + 1/2 = 1$.

On the other hand, if the two spins are antiparallel, their resultant vanishes and $\mathrm{F(A)} = 0$. Thus A_2 yields two composite particles having $\mathrm{F(A)} = 1$ and $\mathrm{F(A)} = 0$ respectively.

The component of the resultant spin angular momentum, $\boldsymbol{F}(\mathrm{A})$, along an external field H_0 will again assume only discrete values in a stationary state. The possible values are characterized by the magnetic quantum number M_A which can assume $2\mathrm{F(A)} + 1$ possible values viz.

$$\begin{aligned} M_A &= \mathrm{F(A)} \\ M_A &= \mathrm{F(A)} - 1 \\ &\vdots \\ M_A &= -\mathrm{F(A)}\,. \end{aligned} \qquad (109)$$

Thus, when $n = 2$, we obtain the value $M_A = +1$, 0 and -1 for the composite particle having $\mathrm{F(A)} = 1$, and $M_A = 0$ for the composite particle having $\mathrm{F(A)} = 0$, or altogether four possible states, *i.e.* the same number of states as obtained by considering the possible combinations of magnetic quantum numbers for the individual nuclei.

The possible values of the resultant spin quantum number F(A) for a system of three magnetically equivalent spins A_3 may be obtained in the following way. First one forms the resultant spin quantum number F(1, 2) for one pair of spins, *1* and *2*, say, to obtain $\mathrm{F}(1,2) = 1$ and $\mathrm{F}(1,2) = 0$, just as in the A_2 group. Thereafter one forms the resultant of $\boldsymbol{F}(1,2)$ and the third spin $\boldsymbol{I}(3)$. Now, if $\mathrm{F}(1,2) = 1$, the resultant $\mathrm{F(A)} = \mathrm{F}(1,2,3)$ will assume the magnitude of $\mathrm{F(A)} = 3/2$ if $\boldsymbol{I}(3)$ is parallel with $\boldsymbol{F}(1,2)$, and $\mathrm{F(A)} = 1/2$ if $\boldsymbol{I}(3)$ is opposed to $\boldsymbol{F}(1,2)$. If $\mathrm{F}(1,2) = 0$, the resultant of $\boldsymbol{F}(1,2)$ and $\boldsymbol{I}(3)$ will equal $\boldsymbol{I}(3)$ in magnitude, and we obtain once more a resultant spin quantum number of $\mathrm{F(A)} = 1/2$. Thus, A_3 yields *one* composite particle with $\mathrm{F(A)} = 3/2$ and *two* composite particles with

F(A) = 1/2. With F(A) = 3/2 one obtains the four magnetic substates of $M_A = 3/2$, 1/2, −1/2, −3/2, and with F(A) = 1/2 one obtains the two magnetic substates $M_A = +1/2$ and $M_A = -1/2$. Altogether this yields one magnetic substate with $M_A = 3/2$, three magnetic substates with $M_A = 1/2$ (one of which belongs to F(A) = 3/2, while the two remaining ones belong to the two combinations having F(A) = 1/2), three magnetic substates with $M_A = -1/2$ and one having $M_A = -3/2$. Thus one obtains altogether 8 magnetic substates, *i.e.* the same number of states as attained by considering the possible combinations of magnetic quantum numbers for the individual nuclei.

Table 5. *Resultant spin states for A_n groupings*

n	F(A)	η[a]	2F(A)+1	η(2F(A)+1)	
1	1/2	1	2	2	
2	1	1	3	3	4
	0	1	1	1	
3	3/2	1	4	4	8
	1/2	2	2	4	
4	2	1	5	5	16
	1	3	3	9	
	0	2	1	2	
5	5/2	1	6	6	32
	3/2	4	4	16	
	1/2	5	2	10	
6	3	1	7	7	64
	2	5	5	25	
	1	9	3	27	
	0	5	1	5	

[a] η represents the number of independent vectorial combinations of the n spins that yield the same value of F(A).

By the same argument, one may show that the resultant spin quantum numbers F(A) of an A_4 grouping are F(A) = 2 (one combination only), F(A) = 1 (three different combinations) and F(A) = 0 (two different combinations). These results are summarized in Table 5, and the combination method is illustrated in Fig. 39.

In spectra containing a group of magnetically equivalent nuclei A_n, the interaction of the spins of the A_n group with the external magnetic field H_0 and with a spin-coupled nucleus B may be represented entirely in terms of the resultant spin angular momentum vector $\boldsymbol{F}$(A). Let us consider the interaction of the nuclei with the H_0 field. Since the Larmor frequencies of the nuclei in the group are all equal, the interaction energy may be written

$$\nu_A I_z(1) + \nu_A I_z(2) + \ldots \nu_A I_z(n) = \nu_A F_z(A)\,. \tag{110}$$

Similarly, the spin-spin coupling between any nucleus in the group and a nucleus B outside the group is one and the same for all the nuclei and may be written J_{AB},

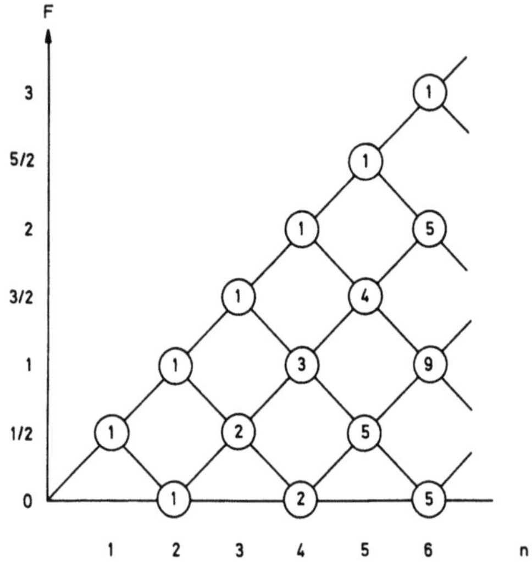

Fig. 39. Total spin quantum numbers F and their statistical weights for n equivalent spin 1/2 nuclei

so that the total spin-coupling interaction between the A nuclei and the B nucleus may be written

$$J_{AB}\boldsymbol{I}(1)\cdot\boldsymbol{I}(B)+J_{AB}\boldsymbol{I}(2)\cdot\boldsymbol{I}(B)+\cdots+J_{AB}\boldsymbol{I}(n)\cdot\boldsymbol{I}(B)=J_{AB}\boldsymbol{F}(A)\cdot\boldsymbol{I}(B)\,. \qquad (111)$$

Thus, for any given resultant spin angular momentum vector $\boldsymbol{F}(A)$, the interaction of the A_n group with external agents may be viewed as an interaction between a composite particle of spin F(A) and the external agent. What the resultant spin quantum number F(A) represents is in fact the internal structure of the A_n group – as long as the *n* A-nuclei remain magnetically equivalent, this internal structure will remain invariant with time. Technically this is expressed by the statement that F(A) is a good quantum number.

b) Subspectra Based on the Composite Particle Concept

The spectra of a spin system that includes a group A_n of magnetically equivalent nuclei may be decomposed into subspectra according to the resultant spin quantum number F(A) of the A_n group, because transitions involving a change in F(A) are strictly forbidden[11]. The procedure to be followed will first be illustrated with a trivial example of a first order A_2X spectrum and it will be shown that the superposition of the subspectra produces the well-known first order multiplets.

[11] It will be noted that this decomposition into subspectra is more fundamental than that based on the concept of effective Larmor frequencies, because *every* transition in the spectrum may be assigned a definite value of F(A). On the other hand, the effective Larmor frequency approach, as used for example in the ABX case, is based on the fact that the magnetic quantum number M_X of the weakly coupled group X may be specified for every transition in the AB basic multiplet, but the transitions in the X basic multiplet may *not* be so classified.

To obtain the correct results, one must take into account that the intensities depend both on the level populations and on the probabilities for transition among levels. Now the level populations are very easily obtained because the Boltzmann factor of every magnetic substate is very close to unity. Thus, a spin system comprising N nuclei will have a population of 2^{-N} for every stationary state.

Secondly, one must note that the transition probability in which the magnetic quantum number, M_A, of the composite particle having $F(A)=F$ changes from M to M+1 depends on F and M and is given by the expression

$$2(F-M)(F+M+1)\,. \tag{112}$$

The expression (112) is valid also for the case when the "composite particle" is just a single nucleus with $F=I=1/2$ (and $M=-1/2$, since the transition is between M and M+1). The factor of two multiplying the brackets in Eq. (112) has been inserted for normalization purposes. Since the fraction of spin 1/2 nuclei which, before the transition, existed in the state $M=-1/2$ is very close to 0.5, only one half of the nuclei in the sample are capable of undergoing

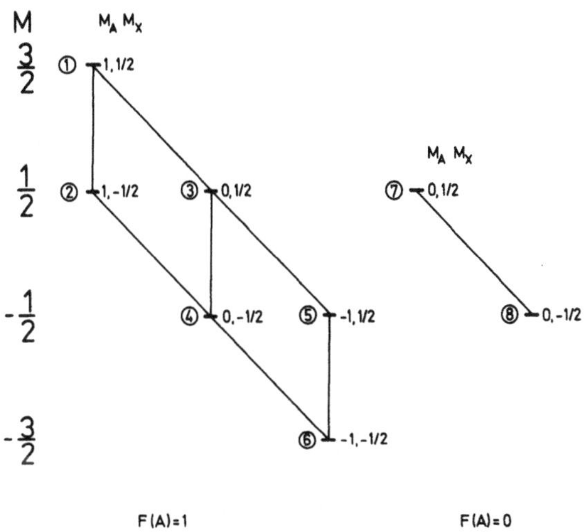

Fig. 40. Energy-level and transition diagram for an A_2X spin system. The energy-level diagram has been decomposed into two parts corresponding to the two different values of the total spin quantum number F of the A_2 group. The pair of numbers next to each level shows the magnetic quantum numbers M_A and M_X in the different states

a transition from M to M+1. Thus one obtains a normalized intensity of *one unit per nucleus* in the sample if the intensity is estimated simply by multiplying the transition probability with the number of nuclei available for making the transition. The energy-level and transition diagram of an A_2X system is shown in Fig. 40, where the decomposition of the spectrum according to F(A) has been displayed explicitly. In this mode of display every energy level corresponds to one single state of the spin system, so that the number of molecules in any one energy level is very close to 1/8 of the total. Since the A_2 and X nuclei are weakly coupled, the levels may be labelled by the magnetic quantum numbers M_A and M_X of the two basic groups. In the diagram of Fig. 40, the vertical lines represent transitions in the X basic multiplet. The decomposition of the A_2 and X basic multiplets into subspectra is indicated in Fig. 41. The subspectrum in the X basic multiplet appertaining to $F(A)=0$ corresponds to one single transition, *viz.* that between levels 8 and 7. The transition probability of this line as evaluated by use of Eq. (112) (with quantum numbers $F=1/2$ and $M=-1/2$ appropriate to the X nucleus) equals 2, and the relative occupation number of level 8 is 1/8 so that this line has an intensity corresponding to 1/4 nucleus. The transition frequency of this line is given simply by the Larmor frequency of nucleus X, since no spin coupling energy arises in molecules having $F(A)=0$ (cf. Eq. (111)). The subspectrum is reproduced in Fig. 41a.

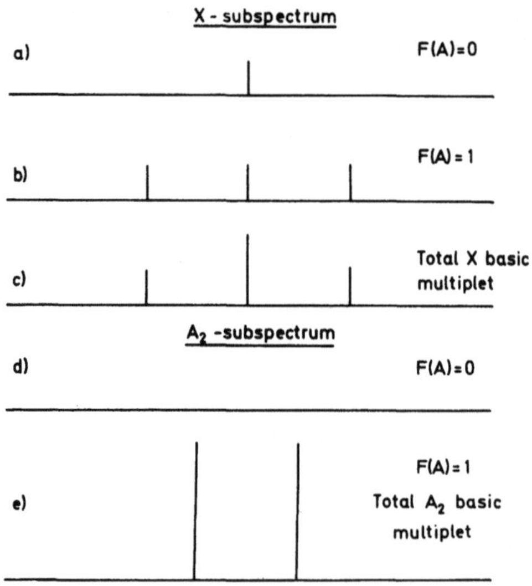

Fig. 41. The A_2X spectrum decomposed according to the two total spin quantum numbers F of the A_2 group. The total X basic multiplet is shown in diagram c) while the total A_2 basic multiplet is shown in diagram e)

The second subspectrum in the X basic multiplet originates in molecules having $F(A)=1$ and consists of three lines corresponding to transitions between the pairs of levels 6–5, 4–3, and 2–1 of Fig. 40. Each one of these lines will again have an intensity corresponding to 1/4 nucleus, so that the intensities of the four lines of the two subspectra add up to one full nucleus. The frequency of the transition 4–3 again equals ν_X (since the product $\boldsymbol{F}(A)\cdot\boldsymbol{I}(X)$ in Eq. (111) simplifies into $M_A \cdot M_X$ in the weakly coupled A_2X spin system), and the frequencies of transition 6–5 and 2–1 will be displaced by $\pm J_{AX}$ from ν_X. In this way one obtains the X subspectrum depicted in Fig. 41b. The total X basic multiplet is a superposition of the subspectra (a) and (b) in Fig. 41 and is depicted in Fig. 41c as the well-known 1:2:1 triplet of the X part of an A_2X spectrum.

Molecules having F(A) = 0 do not contribute to the spectrum of the A_2 basic multiplet, since M_A cannot change from its value of $M_A = 0$ when F(A) = 0. With F(A) = 1 one obtains four transitions in the A_2 basic multiplet, *viz.* those between the pairs of levels 6–4, 4–2, 5–3 and 3–1. The transition probabilities of these lines as evaluated by Eq. (112) are all equal to 4. Since any one intitial state of these transitions is populated by 1/8 of the molecules available, the combined intensity of these lines correspond to two nuclei. The resonance frequencies of the transitions 6–4, 4–2, which originate in molecules having $M_X = -1/2$, are equal and given by $\nu_A - J_{AX}/2$, and the transitions 5–3 and 3–1 coincide at a frequency of $\nu_A + J_{AX}/2$. Thus, the entire A_2 basic multiplet is derived from the subspectrum appertaining to F(A) = 1 (Fig. 41d and e). It is evident, then, that the spectrum as calculated by superposition of subspectra based on the total spin quantum number F(A) is identical with that calculated from the simple rules of Section IV.1.

IV.7. Spectra Characterized by One Spin-coupling Constant: A_nB_m

a) The Structure of A_nB_m Spectra

For a given spin system of the A_nB_m type, the appearance of the spectrum depend on three parameters, ν_A, ν_B and J_{AB}. The intensities in the spectrum depend only on the magnitude of $J_{AB}/(\nu_A - \nu_B)$ and, in fact, the whole spectrum may be characterized in terms of this single parameter. For, if one calculates a series of theoretical spectra all having $\nu_A = 1$ and $\nu_B = 0$ and with J_{AB} ranging from zero to infinity, then one of these spectra will match any given spectrum, apart from a scale factor (positive or negative) to match the actual value of $\nu_A - \nu_B$, and a rigid displacement to match the actual value of ν_B. Such tables of calculated spectra have in fact been compiled and are listed in Appendix B. In the theoretical calculations and also in the classification of the different transitions in A_nB_m spectra, the decomposition of the spectrum into subspectra according to the total spin quantum number of the composite particles will be employed. This decomposition is displayed in Figs. 42 and 43. The total number of stationary states in an A_nB_m spin system is given by $2^{(n+m)}$ when I(A) = I(B) = 1/2. In Figs. 42 and 43 the energy-level and transition diagrams have been decomposed into subspectra based on total spin, with the possible combinations of F(A) and F(B) being obtained from Table 5 or from the diagram in Fig. 39. The simplest nontrivial example is A_2B with F(B) = 1/2 and F(A) = 1 or F(A) = 0, yielding two subspectra with two and six energy levels in the respective subspectral diagrams. The numbers in square brackets represent the appropriate values of F(A) and F(B) in that order. The numbers in ordinary brackets represent the total intensity of all the transitions within a subspectrum normalized to unit intensity for a single nucleus. This total intensity is an invariant, independent of the extent of coupling, and may be calculated by reverting to the weakly coupled limit and using the approach of Section IV.6.b.

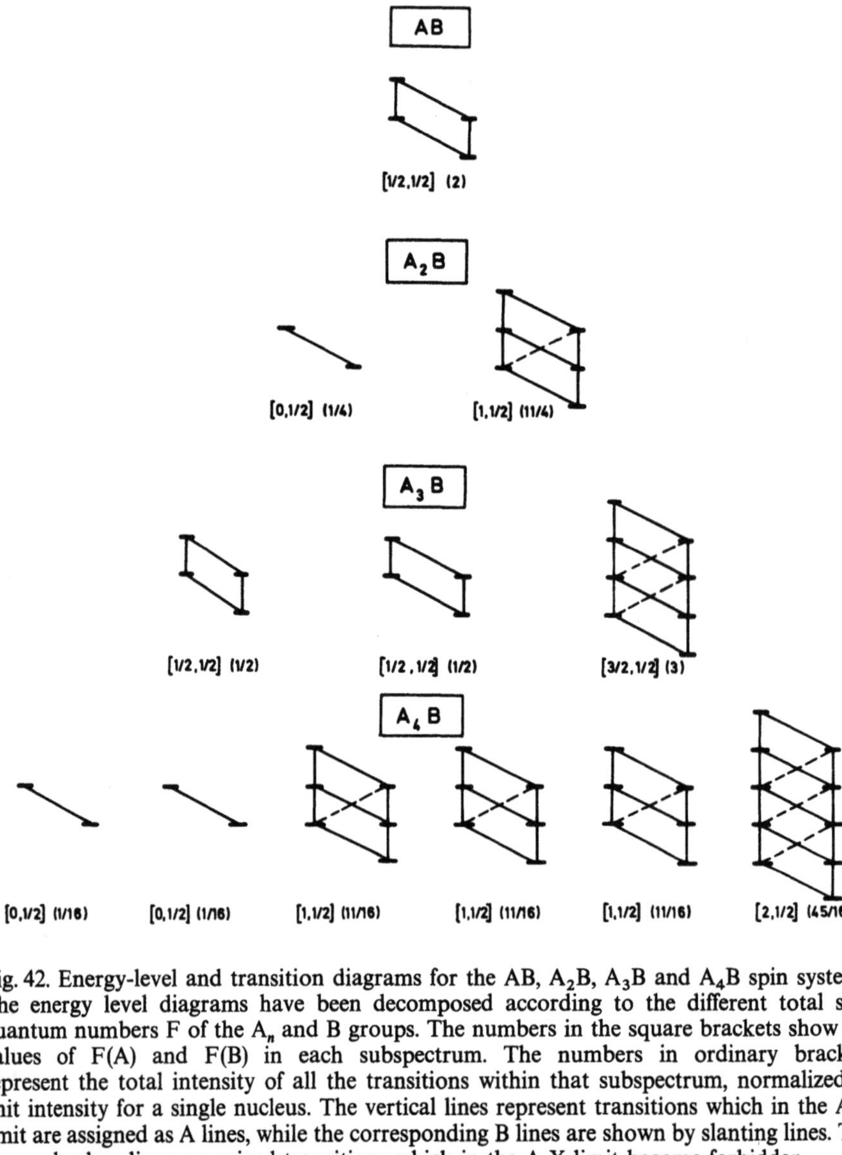

Fig. 42. Energy-level and transition diagrams for the AB, A_2B, A_3B and A_4B spin systems. The energy level diagrams have been decomposed according to the different total spin quantum numbers F of the A_n and B groups. The numbers in the square brackets show the values of F(A) and F(B) in each subspectrum. The numbers in ordinary brackets represent the total intensity of all the transitions within that subspectrum, normalized to unit intensity for a single nucleus. The vertical lines represent transitions which in the A_nX limit are assigned as A lines, while the corresponding B lines are shown by slanting lines. The broken lines are mixed transitions which in the A_nX limit become forbidden

In the transition and energy-level diagrams, the vertical lines represent transitions, which at the limit of weak coupling ($|J_{AB}| \ll |\nu_A - \nu_B|$) are assigned to transitions of the A nuclei, the full lines slanting downwards to the right represent the corresponding B transitions, and the slanting dashed lines represent the so-called combination transitions which become forbidden in the limits of very weak or very strong coupling and which – quite generally – tend to remain unobservably weak in A_nB_m spectra. In the subspectrum [0, 1/2] of the A_2B system the sole transition must have a frequency of ν_B, since $F(A) = 0$

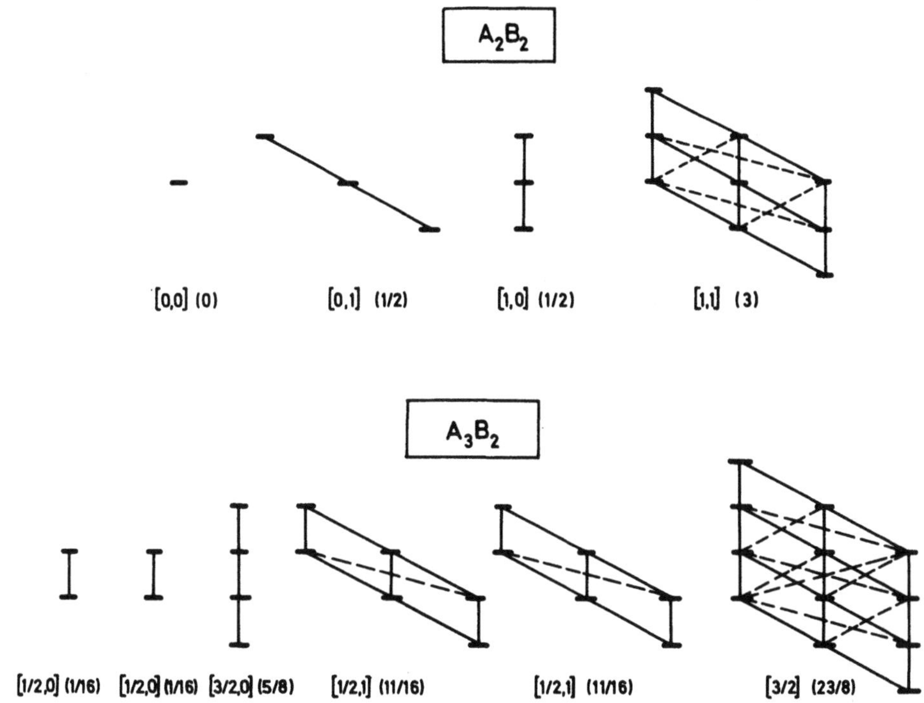

Fig. 43. Same as Fig. 42 for the A_2B_2 and A_3B_2 spin systems

cannot yield any spin coupling effects (cf. Eq. (111)). This transition has an intensity of 1/4 nucleus but at the limit of weak coupling it coincides with one of the transitions from the [1, 1/2] subspectrum.

Similarly, the A_4B and the A_2B_2 spin systems – or, more generally, any A_nB_m spin system with an even n – produce subspectra for which $F(A)=0$ and hence contain a resonance line centered at ν_B, irrespective of the value of J_{AB}. Conversely, A_nB_m spin systems with even values of m, such as A_2B_2 and A_3B_2, contain subspectra with $F(B)=0$ and hence yield a transition centered at ν_A. In other words: an A_nB_m spin system will yield subspectra of the a_n type if m is even, and subspectra of the b_m type if n is even, but these subspectra will not in general have an intensity corresponding to an integral number of nuclei.

The structure of the energy-level and transition diagrams reveals the existence of additional subspectra that resemble spectra of simple spin types. For example, the two [1/2, 1/2] subdiagrams of the A_3B spin system correspond to two coincident *ab* subspectra. The four lines of these subspectra have all the characteristics of a genuine AB spectrum with the true values of ν_A, ν_B and J_{AB}, apart from the integrated intensity which corresponds to *one* nucleus instead of two. Similarly, the three coinciding [1, 1/2] subspectra of the A_4B spectrum have the same transition frequencies and relative intensities as the [1, 1/2] subspectrum of an A_2B spin system with the same values of ν_A, ν_B and J_{AB}. These relationships are sometimes helpful in the analysis procedure.

b) The A_2B Spin System

In Fig. 42, altogether nine transitions have been indicated for an A_2B spin system. In practice, the number of observed transitions is always less because the combination transition is always very weak and at the strong coupling limit two additional transitions approach a vanishing intensity. Furthermore, a number of transitions are degenerate at either the strong coupling or the weak coupling limit.

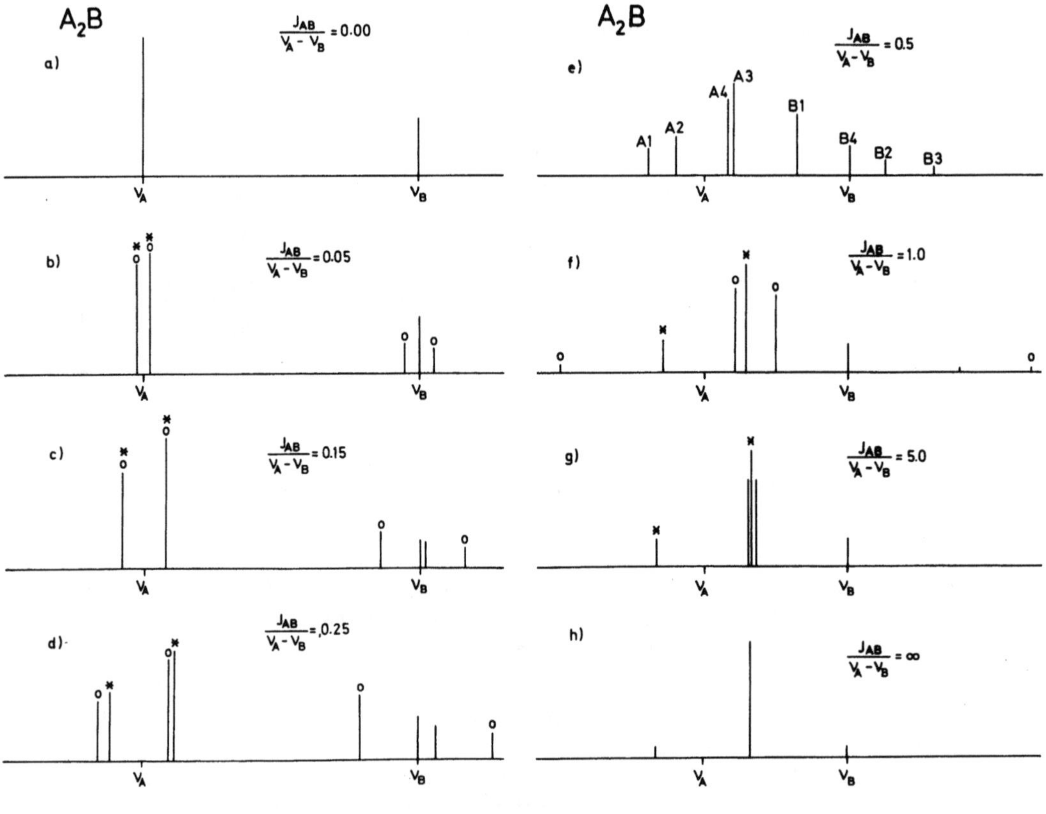

Fig. 44. Theoretical A_2B spectra calculated for a number of different values of $J_{AB}/(\nu_A - \nu_B)$. The A2 and A3 lines used in evaluating ν_A have been marked by a star in spectra b)–d) and f)–g), while the transitions A1, A4, B1 and B3, used in evaluating $|J_{AB}|$ have been marked by circles in spectra b)–d) and f)

A general analysis procedure in the case of A_nB_m spectra is to compare observed frequencies and relative intensities with calculated spectra for a best fit. The diagrams of Fig. 44 are helpful for visual estimate of $|J_{AB}/(\nu_A - \nu_B)|$. These diagrams have all been drawn with a fixed value of $\nu_A - \nu_B$ and with J_{AB} ranging from 0 to ∞. When $|J_{AB}/(\nu_A - \nu_B)| \simeq 0.5$, the chances are usually best that all eight fundamental A_2B transitions can be resolved and the numbering of the lines has been indicated for that particular case. If these eight lines have been identified, the Larmor frequencies and the spin-coupling constant may alternatively be derived from the relationships

$$\nu_B = \nu(B4)\,, \tag{113}$$

$$\nu_A = [\nu(A2) + \nu(A3)]/2\,, \tag{114}$$

$$|J_{AB}| = [\nu(A1) - \nu(A4) + \nu(B1) - \nu(B3)]/3\,. \tag{115}$$

At the limit of very large ratio $|J_{AB}/(\nu_A - \nu_B)|$, the spectrum collapses into a symmetrical triplet with relative intensities 1:10:1 and with a total splitting of $4\,|\nu_A - \nu_B|/3$. In practice, this limit is seldom approached before $|\nu_A - \nu_B|$ falls below the resolving power of the intrument, and the observed spectrum will then consist of one single line.

c) The A_3B Spin System

A series of calculated A_3B spectra is displayed in Fig. 45 and the numbering of the transitions is indicated in the diagram corresponding to $|J_{AB}/(\nu_A - \nu_B)| = 0.5$. The *ab* subspectrum obtained for F(A) = F(B) = 1/2 is composed of lines A7, A8, B5, and B6, respectively, and, if these lines can be identified, the NMR parameters are most readily obtained by an AB analysis of that subspectrum. Otherwise one will have to follow the general A_nB_m analysis procedure of seeking a best fit between calculated and observed spectra.

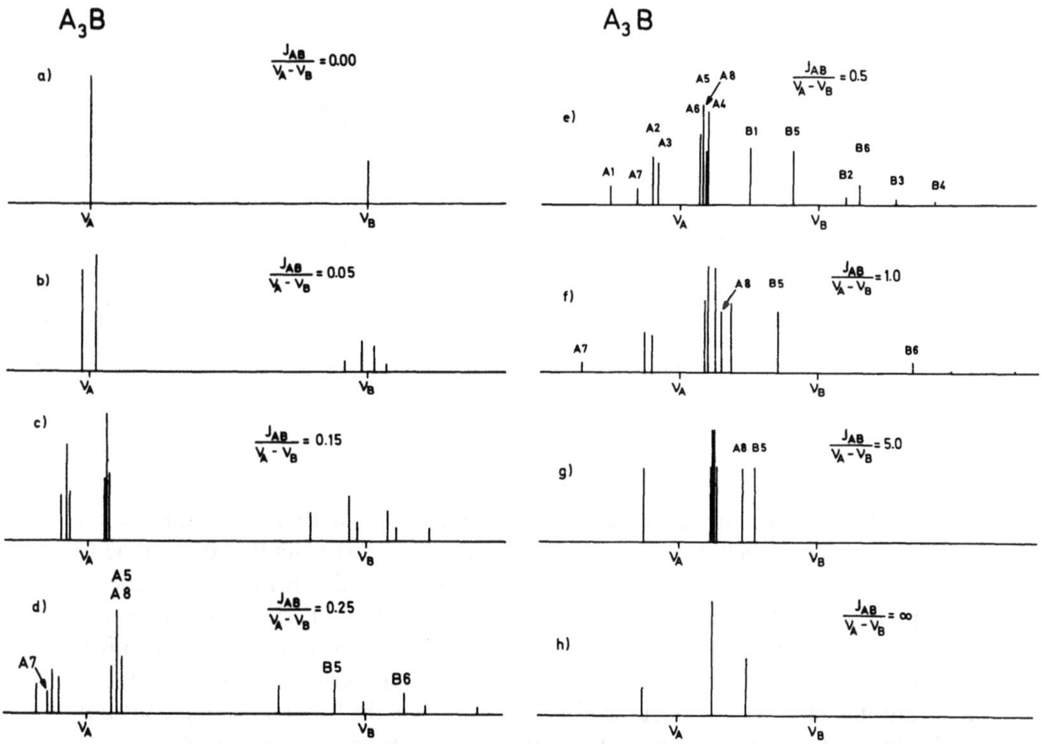

Fig. 45. Theoretical A_3B spectra

d) The A_4B Spin System

The transition and energy level subdiagrams of the A_4B spin system display some resemblance to those of the A_2B spin system. In the actual analysis procedure, little use can be made of this resemblance, however, because the assignment of the observed transitions is seldom obvious. The normal A_nB_m analysis procedure employing comparison between observed and calculated spectra is therefore recommended. A series of calculated A_4B spectra is depicted in Fig. 46. This type of spin system is seldom encountered in organic compounds.

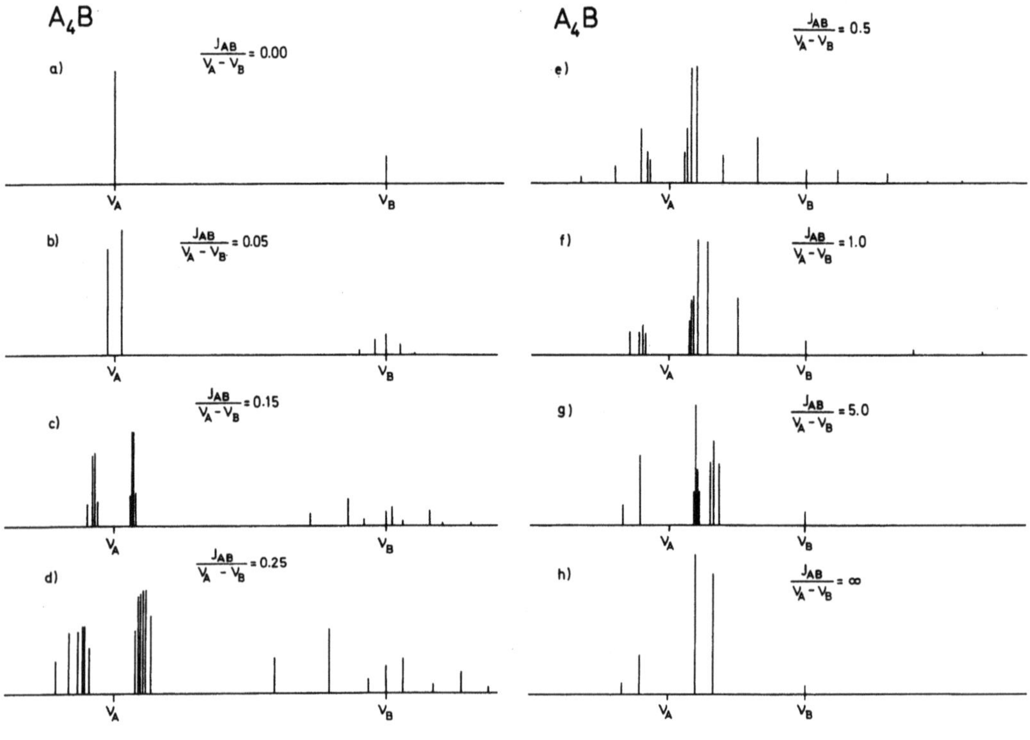

Fig. 46. Theoretical A_4B spectra

e) The A_2B_2 Spin System

The A_2B_2 spin systems usually encountered are in actual fact AA'BB' spin systems which have become deceptively simple, owing to the fact that the magnetic non-equivalence in the latter type of spin system depends on the finite value of one single parameter, *viz.* $L = J_{AB} - J_{A'B}$. For small values of L, deviations from the true A_2B_2 spectrum are of second order in L and tend to become unobservable. A class of compounds in which such spectra might arise is that of the 1,2-disubstituted ethanes. As discussed in Section IV.3.b, the "spin-coupling constant J_{AB}" obtained in such deceptively simple A_2B_2 spectra represents in fact the value of N/2, where $N = (J_{AB} + J_{A'B})$.

The series of calculated A_2B_2 spectra is depicted in Fig. 47. It is seen from the figure that the two lines occurring at ν_A and ν_B are quite prominent in all spectra so that the chemical shifts ν_A and ν_B are readily obtained. The spin-coupling constant is also readily obtained and equals one half of the splitting between lines A1 and A6 and B1 and B6. Unless the spins are very strongly coupled, these lines should also be easily identified. The four last-mentioned transitions in fact yield an *ab* subspectrum with an apparent chemical shift equal to $\nu_A - \nu_B$

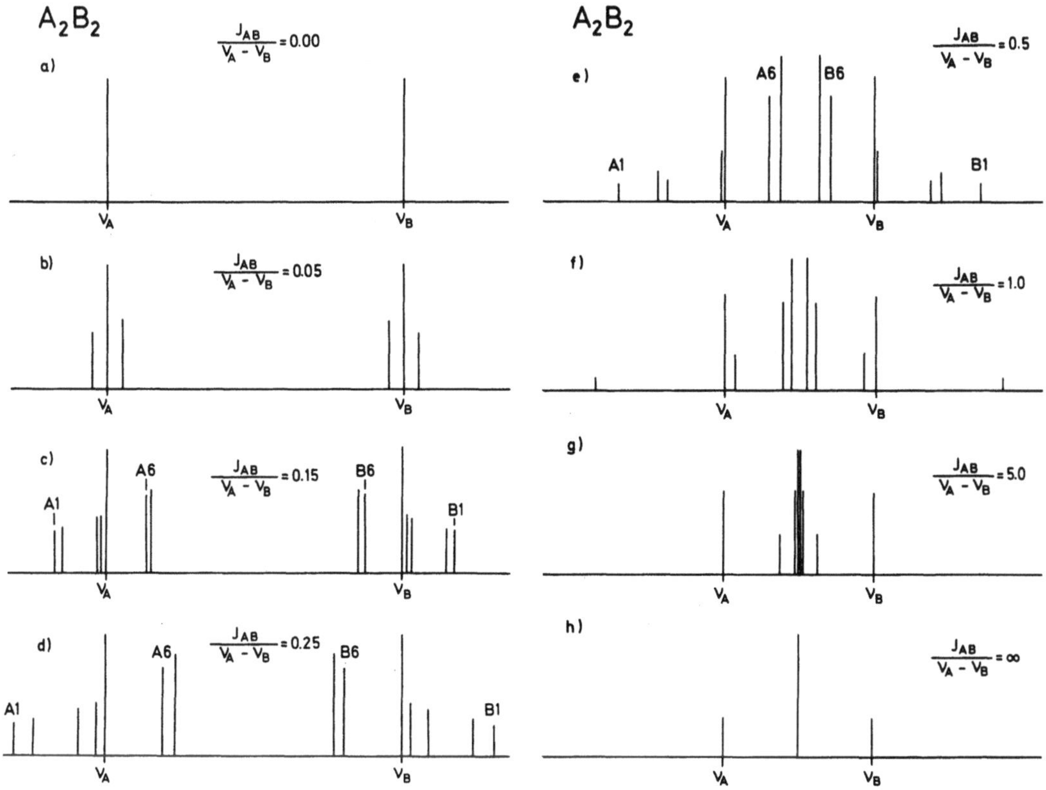

Fig. 47. Theoretical A_2B_2 spectra

and an apparent spin coupling constant equal to $|2J_{AB}|$, as may be deduced from the explicit theoretical expressions. The causes for the occurrence of this *ab* subspectrum are more subtle than those of the other subspectra discussed and are not immediately apparent from the energy and transition diagram of Fig. 43.

f) The A_3B_2 Spin System

This spin system occurs quite frequently in organic compounds containing ethyl groups (cf. the spectrum of ethylbromide in Fig. 21). From the transition and energy-level diagram depicted in Fig. 43, it is obvious that the spectrum includes an a_3 subspectrum and two identical ab_2 subspectra. These subspectra together yield a strong peak at ν_A which is usually readily identified. The position of ν_B may be obtained as the midpoint between transitions B10 and B11 (Fig. 48). To obtain the value of J_{AB}, one may employ the method of Section IV.1.b if the spins are comparatively weakly coupled, but otherwise comparison with theoretical spectra is recommended.

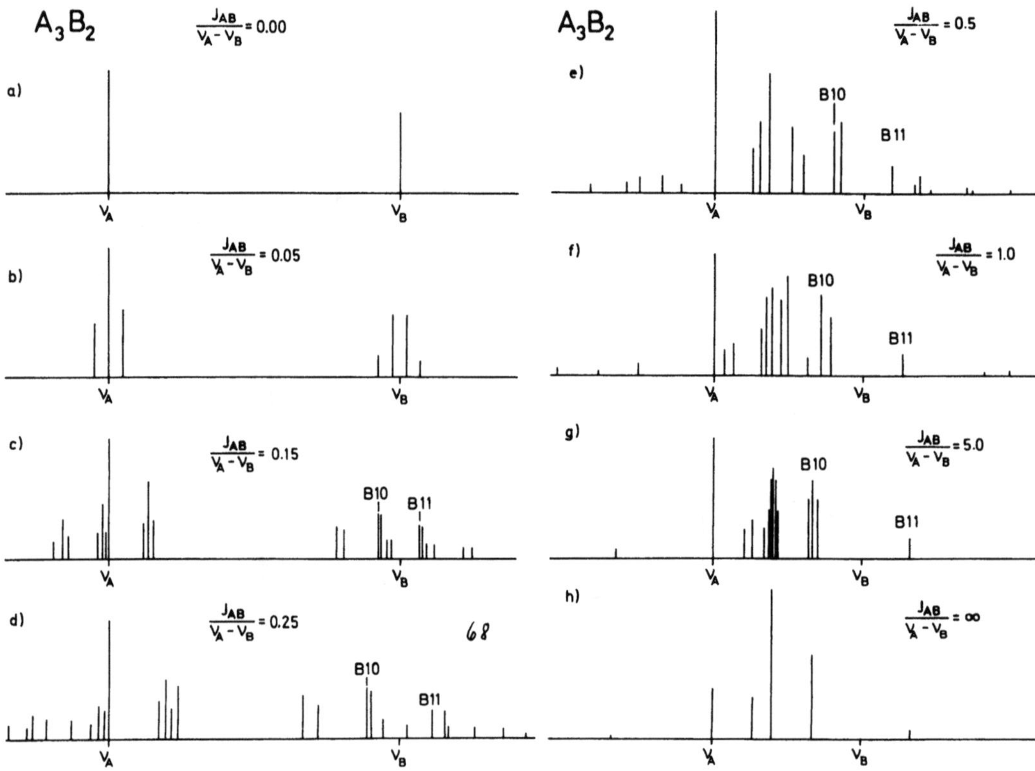

Fig. 48. Theoretical A_3B_2 spectra

Example 1: The A_2B Spectrum of 1,2,3-trichlorobenzene

The 56.4 MHz spectrum of 1,2,3-trichlorobenzene in $CDCl_3$ solution (10% by weight) is shown in Fig. 49. From the symmetry of the molecule, it is clear that the shifts of protons H4 and H6 are the same as are their spin couplings to proton H5. The proton resonance spectrum should thus be of the A_2B type, the appearance of which is characterized by the ratio $|J_{AB}/(\nu_A - \nu_B)|$. A comparison with the theoretical spectra calculated in Fig. 44 shows that the experimental spectrum roughly agrees with that calculated, with the ratio $|J_{AB}/(\nu_A - \nu_B)| = 0.5$. According to Section IV.7.b, the NMR parameters ν_A, ν_B and J_{AB} may be evaluated from the line positions in a simple manner by means of Eqs. (113) to (115). The labelling of the experimental lines given in Fig. 49 follows that of

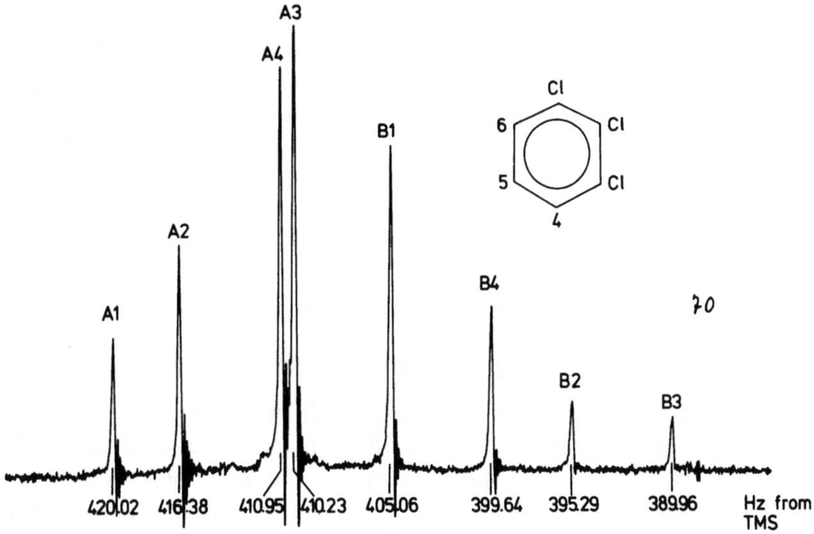

Fig. 49. The NMR spectrum at 56.4 MHz of 1,2,3-trichlorobenzene in a $CDCl_3$ solution (10% by weight). The numbering of the transitions is the same as in Fig. 44

the schematic spectrum in Fig. 44. From Eq. (113) we obtain

$$\nu_B = 399.64 \text{ Hz}.$$

From Eq. (114) we get

$$\nu_A = 413.31 \text{ Hz}$$

and, finally, from Eq. (115) we find

$$|J_{AB}| = 8.06 \text{ Hz}.$$

From other sources it is known that the sign of J_{AB} is positive.

IV.8. The Structure of $A_nB_mX_p$ Spectra

The A_nB_m basic multiplet of an $A_nB_mX_p$ spectrum may be decomposed into $p+1$ subspectra corresponding to the $p+1$ possible values of M_X (assuming $I(X)=1/2$). These subspectra will be of the a_nb_m type with the effective Larmor frequencies given by

$$\nu_a = \nu_A + J_{AX}M_X \tag{116}$$

and

$$\nu_b = \nu_B + J_{BX}M_X. \tag{117}$$

The main difficulty in analysing the A_nB_m basic multiplet consists in identifying the different subspectra. Once this identification has been achieved, the effective Larmor frequencies of the various subspectra and the effective spin-coupling constants – which all equal J_{AB} – of the subspectra are evaluated by the methods described in the preceding section. From this analysis one may extract the values of ν_A, ν_B, J_{AB}, J_{AX}, and J_{BX} and the relative sign of the two latter coupling constants. The X basic multiplet in general becomes asymmetric with respect to its center of gravity at ν_X, and the sense of this asymmetry yields the sign of J_{AB} relative to J_{AX} and J_{BX}. Only the ABX_n spin system is exceptional and yields always a symmetrical X multiplet (cf. Section IV.4 above). The detailed analysis of the X basic multiplet is somewhat involved and the reader is referred to the specialized literature [*11*].

IV.9. The Symmetrical Four-Spin Systems AA′XX′ and AA′BB′

a) The Energy-Level and Transition Diagram

The AA′BB′ and the AA′XX′ spin systems differ primarily in the number of basic groups; their energy-level and transition diagrams may be discussed jointly. Let us consider the AA′BB′ spin system for the sake of concreteness. The symmetrical equivalence implied in the notation means (*i*) that if an interchange of A and A′ is performed concurrently with an interchange of B and B′, symbolically

$$\begin{aligned} A &\leftrightarrow A' \\ B &\leftrightarrow B' \end{aligned} \tag{118}$$

the interaction of the spin system remains invariant and (*ii*) that such a concerted interchange can at most change the signs of the wave function. The invariance of the interactions implies the following constraints on the NMR parameters:

$$\begin{aligned} \nu_A &= \nu_{A'} \\ \nu_B &= \nu_{B'} \\ J_{AB} &= J_{A'B'} \\ J_{AB'} &= J_{A'B}\,. \end{aligned} \tag{119}$$

The symmetrical four-spin system arises naturally in spin system of C_{2v} symmetry but may also arise through motional averaging, as in the case of 1,2-disubstituted ethanes. The spin-spin interactions in the AA′BB′ spin system and their symmetry relations may be represented graphically, as in Fig. 50. The expressions for the transition energies and intensities in AA′BB′ and AA′XX′ spin systems are,

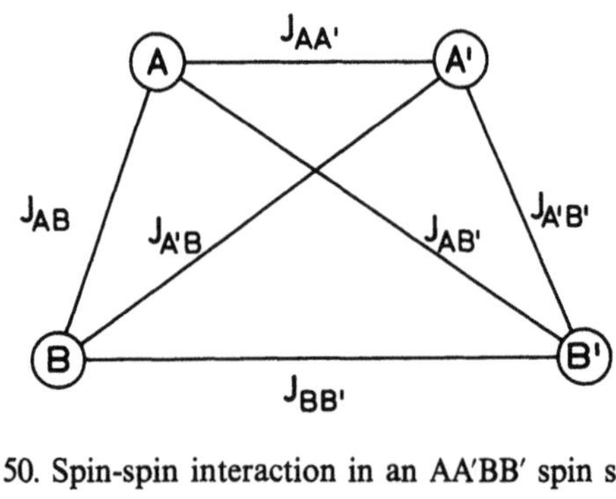

Fig. 50. Spin-spin interaction in an AA′BB′ spin system

however, more conveniently expressed in terms of the following quantities:

$$\begin{aligned} K &= J_{AA'} + J_{BB'} \\ L &= J_{AB} - J_{A'B} \\ M &= J_{AA'} - J_{BB'} \\ N &= J_{AB} + J_{A'B} . \end{aligned} \tag{120}$$

It will be noted that the signs of L and M depend only on the labelling of the spins and hence have no physical significance.

The interaction of the nuclei with the stimulating r.f. field is such that the symmetry character of the state cannot change and the transition and energy-level diagram decomposes into two separate diagrams. Naturally, the spectrum may be similarly decomposed into two subspectra, such that all the transitions within one subspectrum correspond to transitions among stationary states that remain invariant under the symmetry operation of Eq. (118), while the transitions within the other subspectrum correspond to transitions among states whose wavefunctions change in sign under the symmetry operation of Eq. (118). For brevity, these subspectra will be referred to as the *symmetric* and the *antisymmetric* subspectrum respectively, and similarly for transitions and stationary states. The energy-level and transition diagram may be constructed in the following way: from one pair of spin 1/2 nuclei AA' one may construct one antisymmetric state, *viz.* the single state having $M_{AA'} = 0$ and $F(A) = 0$ and three symmetric states having $M_{AA'} = 1$, 0 and -1 respectively and $F(A) = 1$. Similarly, one may construct one antisymmetric and three symmetric states for the BB' pair (Fig. 51).

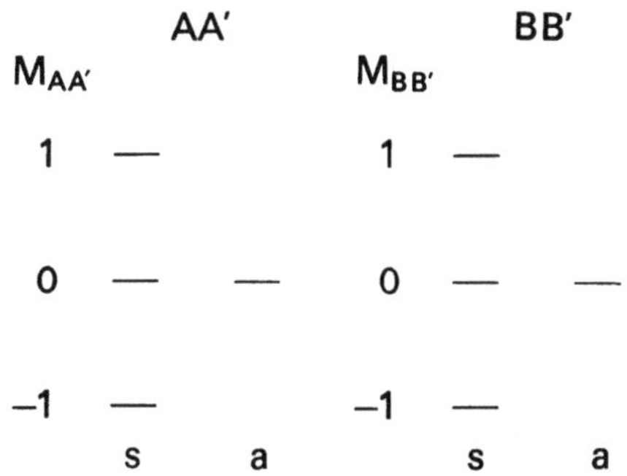

Fig. 51. Symmetric (s) and antisymmetric (a) states in an AA' and a BB' spin system respectively

If there existed no interaction between the AA' and BB' pairs, the states of the combined AA'BB' system would be obtained by constructing all possible combinations of one AA' and one BB' state from the species just considered. When a symmetric AA' state is combined with a symmetric BB' state, the combined AA'BB' state will also be symmetric under the concerted interchange of Eq. (118), and this will also be true for the AA'BB' state obtained by combining the antisymmetric AA' state with the antisymmetric BB' state, because the two sign reversals cancel out. However, if a symmetric AA' state is combined with the antisymmetric BB′ state, or *vice versa*, the combined state comes out antisymmetric. From Fig. 51 it follows that one may construct one symmetric state for the AA′BB′ spin system with a total magnetic quantum number $M = M_{AA'} + M_{BB'} = 2$ and, similarly, one may construct one symmetric state with $M = -2$. For $M = 1$ one obtains two symmetric states, *viz.* by combining a state with $M_{AA'} = 1$ with the symmetric state having $M_{BB'} = 0$, or by combining the state having $M_{BB'} = 1$ with the symmetric state having $M_{AA'} = 0$. Similarly, one obtains two antisymmetric states having $M = 1$ by combining $M_{AA'} = 1$ with the antisymmetric $M_{BB'} = 0$ state, and by combining the $M_{BB'} = 1$ state with the antisymmetric $M_{AA'} = 0$ state. By exactly analogous arguments, one obtains two symmetric and two antisymmetric combined AA′BB′ states with $M = -1$. One may construct four distinct symmetric states all having $M = 0$, three of which are obtained by combining a symmetric AA′ state with a symmetric BB′ state with $M_{AA'} = -M_{BB'}$, the fourth being obtained by combining the two antisymmetric AA′ and BB′ states. No more than two antisymmetric combined states can be obtained that correspond to $M = 0$, and these must both have $M_{AA'} = M_{BB'} = 0$. In this manner one obtains the energy-level diagrams of Fig. 52.

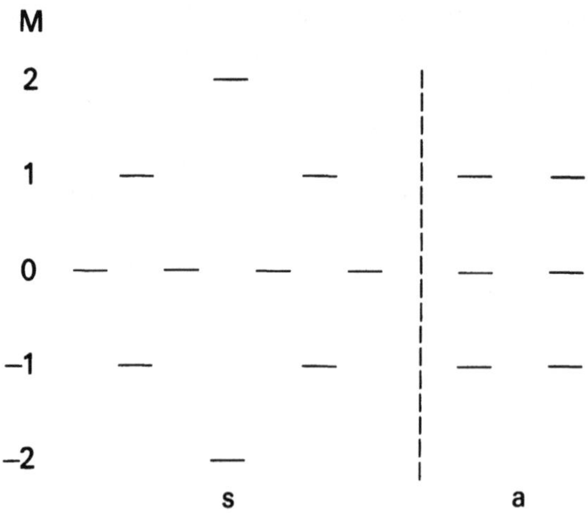

Fig. 52. Symmetric (s) and antisymmetric (a) spin states in an AA′BB′ spin system, obtained by combination of the spin states in Fig. 51

When spin-spin interaction between the AA′ and BB′ pairs of nuclei are included but subject to the constraints of Eq. (119), the stationary states can no longer be classified by giving separately the magnetic quantum numbers $M_{AA'}$ and $M_{BB'}$ (except in the weakly coupled AA′XX′ limit) nor will the stationary states correspond to a definite symmetry type with respect to the AA′ or BB′ species by themselves. However, the number of stationary states belonging to a particular magnetic quantum number M and having a definite symmetry with respect to the concerted interchange of Eq. (118) remains invariant. Thus, the energy-level diagrams of Fig. 51 are valid for any extent of coupling in the AA′BB′ spin system.

The selection rules for the transitions are (*i*) $\Delta M = \pm 1$ and (*ii*) the conservation of symmetry character. Some of the allowed transitions, usually referred to as combination transitions, are always very weak and will not be considered henceforth. The total transition and energy-level diagram for the symmetrical AA'BB' spin system thus appears as in Fig. 53. To the left of each energy level in Fig. 53 appears a spectroscopic label, for example $2s_0$, in which the first number is simply a running label to distinguish the $2s_0$ state from other s_0 states, the letter *s*

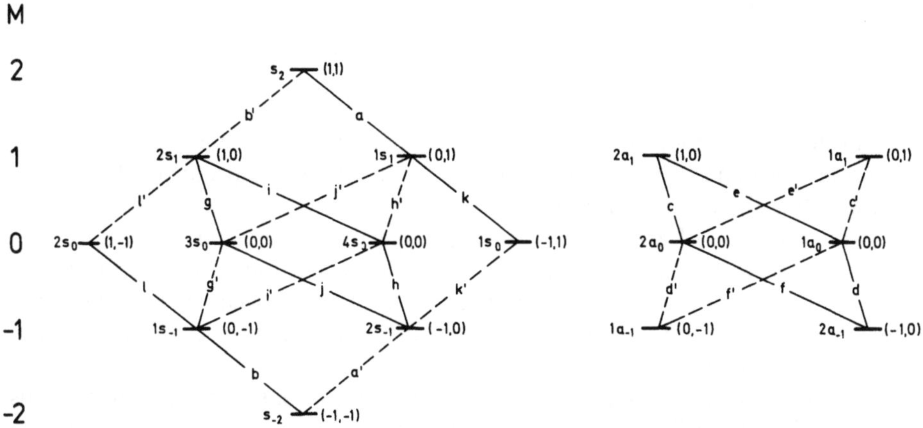

Fig. 53. Total transition and energy-level diagram for an AA'BB' spin system. The states are labelled s for symmetric and a for antisymmetric. The suffix on the state labels gives the total magnetic number M, while the first number is simply a running label to distinguish the states with the same value of M. The numbers in brackets to the right of each level are the magnetic quantum numbers $M_{AA'}$ and $M_{XX'}$ of the basic groups in the AA'XX' limit

represents the symmetry character (*s* = symmetric, *a* = antisymmetric), and the lower suffix 0 stands for the total magnetic quantum number of the state. To the right of each energy level a bracket encloses two numbers that specify the magnetic quantum numbers of the basic groups $M_{AA'}$ and $M_{XX'}$ respectively in the weakly coupled limit. The slanting lines represent the transitions, those indicated by full lines are classified as AA' transitions and by dashed lines as BB' transitions. Every transition is specified by a letter (*a* through *l*) according to a nomenclature developed by DISCHLER *et al.* [*12*]. It may be shown that an AA'BB' spectrum is always strictly symmetrical with respect to its center of gravity at $(\nu_A + \nu_B)/2$, and a transition designated by a primed letter is always the mirror image of the transition labelled with the corresponding unprimed letter. It is *not* generally true, however, that all unprimed transitions (classified as AA' transitions) occur on the same side of the midpoint as ν_A and *vice versa.*

In the limit when $J_{A'B}$ tends to J_{AB} the AA'BB' spin system approaches the A_2B_2 spin system. In this limit, the transitions *g*, *g'* and *j*, *j'* become strictly forbidden, and the level $3s_0$ becomes disconnected from the nine remaining symmetric levels and the symmetric subdiagram decomposes into the [1,1] and [0,0] subdiagrams of the A_2B_2 spin system (Fig. 43). Similarly, the transitions *c*, *c'* and *f*, *f'* become strictly forbidden in the A_2B_2 limit and the antisymmetric subdiagram decomposes into the two subdiagrams [1,0] and [0,1] of the A_2B_2 spin system. Depending upon the values of the parameters $K = J_{AA'} + J_{BB'}$ and $M = J_{AA'} - J_{BB'}$, the magnetic non-equivalence may become unobservable in either the symmetric or antisymmetric subspectrum. When this is the case there will be some transitions in the AA'BB' spectrum that correspond to A_2B_2 transitions and others that have no such correspondence. It is helpful in this context to note that a detailed analysis establishes the antisymmetric subspectrum to be independent of K and the symmetric subspectrum to be independent of M.

b) The AA'XX' Spin System

In the weakly coupled AA'XX' limit the spectrum decomposes into two basic multiplets assigned to the AA' and XX' basic groups respectively. It may be shown that these basic multiplets are both symmetric with respect to their individual centers of gravity at ν_A and ν_X respectively. Since the entire spectrum is symmetric with respect to $(\nu_A + \nu_X)/2$, it follows that the two basic multiplets are identical in appearance and apart from the Larmor frequency of the other basic group the entire information content of the whole spectrum resides within one basic multiplet.

Let us consider the AA' basic multiplet the transitions of which corresponds to the full slanting lines of Fig. 53, *i.e.* those for which $\Delta M_{AA'} = \pm 1$. The only states that become perturbed by spin coupling between the AA' and XX' basic groups are those that belong to the same symmetry species and have identical sets of quantum numbers for $M_{AA'}$ and $M_{XX'}$. Thus, $3s_0$ and $4s_0$ will mix and so will $1a_0$ and $2a_0$. The transitions *a* and *k*, which occur between unmixed states, will coincide at a resonance frequency $\nu_A + N/2$, as in the A_2X_2 spin system, and their combined intensity equals 1/4 of the total AA' basic multiplet intensity. Similarly, lines *b* and *l* will coincide at $\nu_A - N/2$ and have a combined intensity of 1/4 of the total AA' basic multiplet intensity. The remaining half of the AA' basic multiplet intensity is shared among the 8 transitions *c* through *j*, and hence the two former lines will usually be the strongest ones and are readily identified. Their splitting yields the value of N and the center of gravity of the basic multiplet yields ν_A. The magnitudes of the parameter L and M may be extracted from the antisymmetric subspectrum comprising lines *c* through *f*. These four transitions originate from a four-level system having unmixed uppermost and lowermost levels and thus yield an *ab* type subspectrum. This *ab* subspectrum is centered at ν_A and the effective chemical shift $|\nu_a - \nu_b|$ equals L in magnitude and the apparent *ab* spin-coupling constant equals M in magnitude. The total intensity of this *ab* subspectrum equals 1/4 of the total AA' basic multiplet intensity. Similarly, the four transitions *i* through *h* yield an *ab* type subspectrum, with a total intensity equal to 1/4 of the AA' basic multiplet intensity, centered at ν_A with an apparent chemical shift equal to L and an apparent spin coupling equal to K. Unfortunately, there is no way of telling which of the two *ab* type subspectra is which, and hence it is not possible to distinguish between K and M. Consequently the relative signs of $J_{AA'}$ and $J_{XX'}$ remain undetermined. On the other hand, L and N influence the spectrum in different ways and hence these two entities are distinguishable, thus allowing one to determine the relative signs of J_{AX} and $J_{A'X}$.

If either K or M vastly exceeds L in magnitude, the corresponding *ab* subspectrum becomes deceptively simple and collapses into a single line centered at ν_A. If both K and M are large compared to L, the entire AA' basic multiplet becomes deceptively simple and assumes the 1:2:1 triplet pattern of the A_2 basic multiplet of an A_2X_2 spin system. This is, for example, the case in the spectrum of furan depicted in Fig. 27.

c) Analysis of AA'XX' Spectra

The AA' and XX' basic multiplets are identical in appearance and display mirror-image symmetry with respect to ν_A and ν_X respectively. In the general case each basic multiplet displays 10 lines.

Unless the spectrum is deceptively simple, all 10 lines are resolved in which case the two strongest lines are split by $|N| = |J_{AX} + J_{A'X}|$.

Of the 8 remaining lines in the basic multiplet, the two outermost and the two innermost lines together constitute an *ab* type subspectrum, and the four remaining lines constitute a second *ab* type subspectrum. The integrated intensities of these two subspectra are equal and together comprise 1/2 of the total basic multiplet intensity. The apparent chemical shifts and the apparent spin couplings of the two *ab* subspectra are evaluated by the standard procedure of Section IV.2. The two apparent chemical shifts should be the same and equal to $|L| = |J_{AX} - J_{A'X}|$ in magnitude. The two apparent spin-coupling constants of the *ab* subspectra equal $|K| = |J_{AA'} + J_{XX'}|$ and $|M| = |J_{AA'} - J_{XX'}|$ in magnitude. There is no way to distinguish between K and M, and the only sign information obtainable from the analysis is the relative signs of J_{AX} and $J_{A'X}$.

d) The Structure of AA'BB' Spectra

In the AA'BB' spin system there will be mixing of all states that belong to one and the same total magnetic quantum number M and to one and the same symmetry species. For example, all four s_0 states will be modified as compared to the way in which they were constructed in subsection (a) above. Consequently it is not possible to obtain simple analytical expressions for all transition frequencies and intensities in terms of the NMR parameters, nor will one in general be able to extract unique values of all NMR parameters in the most general case. There exists, however, a number of regularities corresponding to invariants and in certain cases these are sufficient for deriving the NMR parameters.

First, it should be noted that the entire spectrum is symmetric with respect to the center of gravity at $(\nu_A + \nu_B)/2$. Secondly, a number of frequency sum rules are evident from the transition and energy level diagrams. These frequency sum rules may be expressed in a shorthand notation by letting the letters *a* through *l* represent not only the labels of the corresponding transitions but also their frequencies with respect to the center of gravity at $(\nu_A + \nu_B)/2$. The four antisymmetric transitions c to f obey the frequency sum rule

$$c+f=d+e\,. \tag{121}$$

Similarly, the four symmetrical transitions g through j obey the frequency sum rule

$$g+j=i+h\,. \tag{122}$$

Finally, by considering the quadrangle involving levels s_2, $2s_1$, $3s_0$ and $1s_1$ and noting that the symmetry of the spectrum leads to the identities $b'=-b$, $g'=-g$, one may derive the frequency sum rule

$$a+b=g+j\,. \tag{123}$$

No frequency sum rule may be constructed for the transitions k and l because the frequencies of k and k' and of l and l' cancel for every closed polygon involving any one of those transitions. Eqs. (122) and (123) may be combined into the expression

$$a+b=g+j=i+h\,. \tag{124}$$

In the process of analysis, one of the things to look for is the two "antisymmetric" pairs cf and de that have a common mean frequency (cf. Eq. (121)), and the three "symmetric" pairs ab, gj and ih, that have a common mean frequency. (The occurrence of a common mean frequency is one recognizable feature that the experimentalist may employ for utilizing frequency sum rules. Alternatively, one may prefer to re-express them as rules of repeated spacings.)

One invariant of an NMR spectrum that is of relevance for the AA'BB' spin system is the sum of the intensities of all transitions within an energy-level diagram that connect a given pair of magnetic quantum numbers M and M − 1. By normalizing the intensity to that of a single uncoupled nucleus as unity and looking for correspondence with the AA'XX' limit, one may derive the following intensity sum rules

$$I_a+I_b=0.5\,, \tag{125}$$

$$I_c+I_e=I_d+I_f=0.25\,, \tag{126}$$

$$I_g+I_h+I_i+I_j+I_k+I_l=1.0\,. \tag{127}$$

(In deriving these frequency sum rules we have utilized the symmetry of the spectrum such that $I_b=I_{b'}$, $I_e=I_{e'}$ etc.) In addition to these intensity sum rules, one may show by use of a quantum mechanical treatment that the intensities of lines a and b are in the ratio

$$I_a/I_b=b/a \tag{128}$$

i.e. the same ratio as obtained in an AB spectrum (cf. Section IV.2.b). In fact, the four transitions a, b, b', and a' constitute an ab type subspectrum which is the counterpart of the ab type subspectrum obtainable in A_2B_2 spectra. In the AA'XX' limit this ab type subspectrum simplifies into an ax type subspectrum. The effective chemical shift parameter of the subspectrum of lines $aa'bb'$ equals the true chemical shift $|\nu_A-\nu_B|$ and the effective spin coupling constant equals $N=J_{AB}+J_{A'B}$ or – in the A_2B_2 case – simply $2J_{AB}$.

e) Analysis of AA'BB' Spectra

An AA'BB' spectrum is always symmetric with respect to its center of gravity at $(\nu_A + \nu_B)/2$. The information obtainable from the spectrum is the magnitude of $|\nu_A - \nu_B|$ and the magnitude of the spin-coupling parameters K, L, M, and N defined in Eq. (120), as well as the relative signs of K and N. It follows from Eq. (120) that the relative signs of all spin-coupling constants may be determined, but it is not possible to correlate these parameters with molecular structure – *in particular, one does not know which one of the spin couplings $J_{AA'}$ and $J_{BB'}$ goes together with the more screened pair of nuclei.*

The first step in the analysis procedure consist of the assignment of the observed lines to the twelve theoretical transitions *a* through *l*. The frequency sum rules of Eqs. (121) to (124) and the intensity sum rules of Eqs. (125) to (127), as well as the intensity ratio of Eq. (128), are helpful in this effort but do not suffice to give an unambiguous assignment. The possible permutations of assignments among the four antisymmetrical transitions may be shown to be physically irrelevant, and one may set once for all a sequence

$$c \geqq d \geqq e \geqq f \tag{129}$$

corresponding to a positive sign of the parameters M and L. One additional constraint may be imposed, this constraint corresponding to an arbitrary choice of absolute sign of the spin-coupling constants. Either *one* of the two following conditions is possible:

$$a \geqq b \tag{130}$$

or

$$g \geqq h\,. \tag{131}$$

The first of the two latter inequalities corresponds to a convention in which N comes out positive, and the second corresponds to a convention that will always yield a positive K. Since the actual spectrum depends on the relative signs of K and N, one is *not* entitled to *impose* the two inequalities of Eqs. (130) and (131) at the same time.

One ambiguity that cannot be lifted by the above constraints is the assignment of lines *k* and *l* since an interchange of the labels of these two transitions will not change any of the above relationships. If the spectrum does not deviate much from that of an AA'XX', one will always have

$$k \simeq a\,, \tag{132}$$

$$I_k \simeq I_a\,, \tag{133}$$

$$l \simeq b\,, \tag{134}$$

$$I_l \simeq I_b\,. \tag{135}$$

When the four last equations apply one can avoid mistaking *k* for *l* and *vice versa.* However, these conditions imply that the midpoint between lines *k* and *l* nearly coincides with that of the three remaining pairs of symmetric transitions, and unfortunately render difficult the distinction of lines *k* and *l* from lines *a* and *b*. It is therefore recommended to supplement the assignment based on frequency and intensity relationships with double resonance and double quantum spectra. In general, it then becomes possible to arrive at an unambiguous assignment.

Once all the lines have been assigned, the NMR parameters are given by the following equations:

$$|\nu_A - \nu_B| = 2\sqrt{ab} = \sqrt{2(ce+df)}\,, \tag{136}$$

$$K = b+g+k-a-h-l = b+i+k-a-j-l\,, \tag{137}$$

$$|L| = \sqrt{(c-e)^2 - M^2} = \sqrt{(d-f)^2 - M^2}\,, \tag{138}$$

$$|M| = (ce-df)/|\nu_A - \nu_B|\,, \tag{139}$$

$$N = a-b = \pm\sqrt{C^2 - (\nu_A - \nu_B)^2} \tag{140}$$

where

$$C = a+b = g+j = h+i\,. \tag{141}$$

In the above expressions the letters a through l represent the frequencies of the corresponding transitions measured with respect to $(\nu_A + \nu_B)/2$ as origin (cf. subsection (d) above).

The above formulas allow one to evaluate the NMR parameters by simple algebraic manipulations of the observed transition frequencies. A weakness of this approach is that it does not make full use of all the information in an observed spectrum.

For example, it may be shown that the frequencies of the symmetric transitions g to l (in particular those of k and l) do depend on the magnitude of the parameter |L|, but this is not evident from the foregoing formulas where |L| is evaluated entirely in terms of the antisymmetric transitions. Furthermore, the formulas will not allow one to evaluate the parameter K if the $3s_0$ level does not mix sufficiently with the other s_0 levels to render at least one of the transitions g and j observable, as may happen, for example, in 1,2-disubstituted ethanes. When this is the case, however, the spectrum can never yield an accurate value of K, but a full quantum mechanical analysis could provide a reasonably good estimate of this parameter. Nevertheless, the formulas given above (originally derived by DISCHLER *et al.* [*12*]) are quite simple and may be used in a number of actual cases.

In many cases the chemical shift in an AA'BB' spectrum is large enough to allow a unique correlation of the observed lines with those of an AA'XX' spectrum (examples of such spectra will be presented in subsection (f) below). The assignment will then be aided by the use of the following second order perturbation results.

The frequencies of *all* the transitions, when compared with their positions in an genuine AA'XX' spectrum, are shifted away from the center at $(\nu_A + \nu_B)/2$. The magnitude of this displacement to a first approximation equals

$$\varepsilon_1 = L^2/4\,|\nu_A - \nu_B| \tag{142}$$

for the four antisymmetric transitions and also for the two symmetric transitions k and l, but equals

$$\varepsilon_2 = N^2/4\,|\nu_A - \nu_B| \tag{143}$$

for the six remaining symmetric transitions. Thus the midpoint of lines *k* and *l* should nearly coincide with the midpoint of the four antisymmetric transitions. The common midpoint of transitions *a, b, g, h, i, j* will be further removed from the center of the entire spectrum at $(\nu_A + \nu_B)/2$ than the common midpoint of the transitions *c, d, e, f, k, l* if $|N| > |L|$, but closer to $(\nu_A + \nu_B)/2$ if $|N| < |L|$. It will be noted that $|N| > |L|$ implies that the two AB coupling constants J_{AB} and $J_{A'B}$ carry the same sign, whereas $|N| < |L|$ implies an opposite sign of these two coupling constants. If one of the couplings, $J_{A'B}$ say, vanishes, N becomes equal to L and the second order corrections to the frequencies produce a rigid displacement of the entire half-spectrum.

f) Typical Examples of AA'BB' Spin Systems in Proton Magnetic Resonance

The recognizable patterns of an AA'BB' half-spectrum are not at all as conspicuous as are those of an AA'XX' half-spectrum, and when confronted with an unknown spectrum one will not immediately recognize where to start. Unfortunately, it is not possible to present a series of typical AA'BB' patterns applicable to all practical cases because the five-dimensional parameter space is far too large to be comprehended. Instead, we shall discuss the appearance of four classes of AA'BB' spectra that have occurred frequently in the NMR literature. Each of the examples to be dealt with below exhibits its own distinct features and is representative of that region of the parameter space to which it belongs.

The compounds to be discussed are

1. 1,2-disubstituted ethanes (where symmetrical equivalence arises either by symmetry or through motional averaging)

$$X{-}CH_2{-}CH_2{-}Y\,,$$

(I)

2. $-CH_2-CH_2-$ groupings in symmetrical non-planar rings such as ethylene sulphite

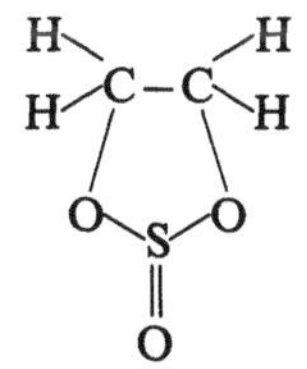

(II)

3. unsymmetrically 1,4-disubstituted benzenes

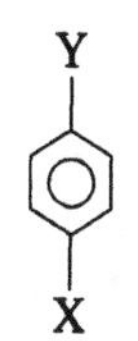

(III)

4. symmetrically 1,2-disubstituted benzenes.

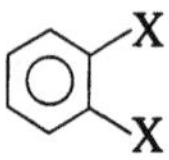

(IV)

The relative signs of spin-coupling constants in compounds of this type have been determined in a number of cases, and it is now generally accepted that all the above proton-proton spin couplings are positive except for the geminal spin couplings in compounds of type (I) and (in most cases) in compounds of type (II). The orders of magnitude of the spin coupling parameters in the four types of compounds are

for (I) $$|K| \gg |N| \gg |L| \simeq |M|, \qquad (144a)$$

for (II) $$|L| \simeq |K| \gg |M| \simeq |N|, \qquad (144b)$$

for (III) $$|N| \simeq |L| > |K| \gg |M|, \qquad (144c)$$

for (IV) $$|N| \geqq |K| \geqq |M| \geqq |L|. \qquad (144d)$$

It will be noted that the parameter $M = J_{AA'} - J_{BB'}$ in compounds of type (I) through (III) may become unobservably small. When this happens, the four antisymmetric transitions *c* to *f* collapse pairwise, *c* and *d* forming one pair and *e*, *f* the other. Together with their counterparts in the other half-spectrum (*c'*, *d'* and *d'*, *f'*), these lines then yield an *ab* type subspectrum the effective chemical shift of which equals $|\nu_A - \nu_B|$ and the effective spin-coupling constant of which equals |L|. More typically, however, M remains finite and has been taken as such in the calculated spectra presented below.

Schematic spectra of a symmetrically 1,2-disubstituted benzene (IV) are reproduced in Fig. 54a–c. Only one half of the symmetrical spectrum is drawn. The spin-coupling parameters assumed in the calculation were

$$\begin{aligned} K &= 8\ \text{Hz} \\ L &= 6\ \text{Hz} \\ M &= 7\ \text{Hz} \\ N &= 10\ \text{Hz}. \end{aligned} \qquad (145)$$

In the upper spectrum (Fig. 54a) the chemical shift $\nu_A - \nu_B$ is assumed to be very large so as to yield an AA′XX′ spectrum, and in the middle spectrum (Fig. 54b) the chemical shift $\nu_A - \nu_B$ was set to 40 Hz, while the lower spectrum (Fig. 54c) has been calculated with $\nu_A - \nu_B = 20$ Hz. In the spectra of Fig. 54a–c the reference frequency has arbitrarily been set equal to ν_A, so that $\nu_A = 0$ and ν_B equals -40 Hz in the middle spectrum and -20 Hz in the lower spectrum. It will be noted that lines *d* and *h* almost coincide in the AA′XX′ limit as do lines *i* and *e*. The reason for this is the near equality of K and M.

As the chemical shift decreases from infinity in spectrum 54a to a large but finite value in spectrum b, the "inner" lines, *b*, *l*, *f*, and *j*, increase in intensity at the expense of the "outer" lines *c*, *g*, *a*, and *k*. With decreasing chemical shift this

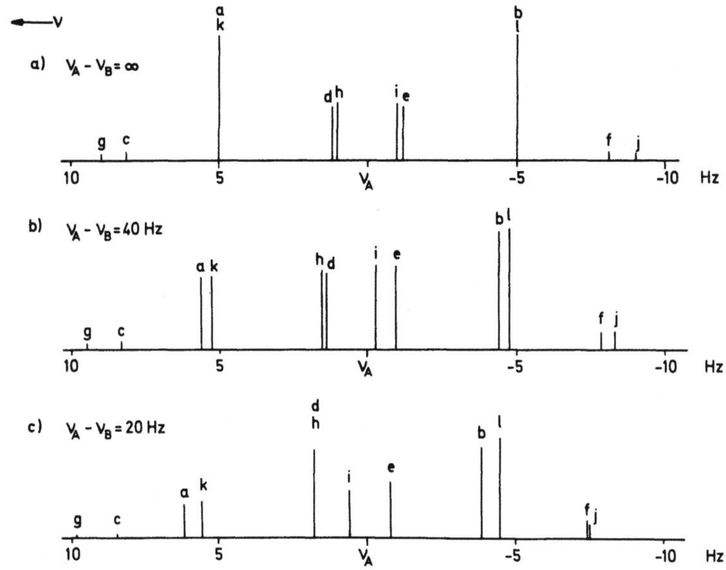

Fig. 54. Low-field half of schematic spectra from a hypothetical symmetrically 1,2-disubstituted benzene, calculated with the parameters K = 8 Hz, L = 6 Hz, M = 7 Hz and N = 10 Hz. The reference frequency has arbitrarily been set equal to ν_A, *i.e.* $\nu_B = -40$ Hz in diagram b) and $\nu_B = -20$ Hz in diagram c)

intensity redistribution continues and lines *g* and *c* may become lost in the background noise. Since $|N| > |L|$, the second order shift (Eqs. (142) and (143)) of lines *a*, *b*, and *g* through *j* is larger than that of the remaining lines and the splitting between lines *g* and *c* tends to become larger than that between lines *f* and *j*. Similarly, the splitting between lines *i* and *e* tends to become larger than that between lines *h* and *d*. This is illustrated in the spectrum in Fig. 54b.

These changes in the line positions continue on further reduction of the relative chemical shift, as can be seen from the spectrum in Fig. 54c, where lines *f* and *j* almost coincide while the splitting between lines *g* and *c* has increased, as has also the splitting between lines *i* and *e*. If the chemical shift is small enough, the lines *f* and *j* of the AA' half-spectrum that are closer to the BB' half-spectrum will cross the corresponding lines *f'* and *j'* of the BB' half-spectrum.

The spectrum of an unsymmetrically 1,2-disubstituted benzene (III) is shown in Fig. 55. As in the previous figure, only the AA' half-spectrum is reproduced. The spin-coupling parameters employed in the calculations were

$$\begin{aligned} K &= 5\ \text{Hz}. \\ L &= 8\ \text{Hz} \\ M &= 1\ \text{Hz} \\ N &= 9\ \text{Hz}. \end{aligned} \qquad (146)$$

In the upper spectrum (Fig. 55a) the chemical shift $\nu_A - \nu_B$ is assumed to be very large and in the middle spectrum (Fig. 55b) the shift $\nu_A - \nu_B$ is set equal to

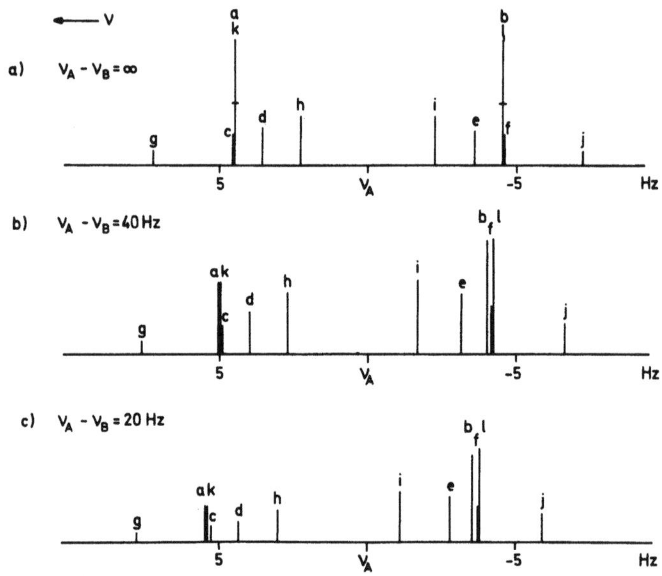

Fig. 55. Low-field half of schematic spectra from a hypothetical unsymmetrically 1,4-disubstituted benzene. The spectra have been calculated with the parameters K = 5 Hz, L = 8 Hz, M = 1 Hz and N = 9 Hz. As in Fig. 54, the reference frequency has arbitrarily been set equal to ν_A

40 Hz while in the lower spectrum (Fig. 55c) $\nu_A - \nu_B = 20$ Hz. Since |L| approximately equals |N|, and because these two parameters are both appreciably larger than |K| and |M|, the half-spectrum decomposes into two distinct multiplets, the dominant lines of which are *a*, *k* and *b*, *l*, respectively. The two antisymmetric transitions *c* and *f* nearly coincide with the two last-mentioned pairs of lines, so that in practical cases at most eight lines can be resolved in each half-spectrum. Lines *d* and *e* will always be close to the two strongest lines of the spectrum, but must be resolved in order to allow a good estimate of the magnitude of |M| to be obtained. The four remaining lines in the half-spectra may be assigned to transitions *g* through *j*. The multiplet splittings within a half-spectrum do not depend very strongly on $\nu_A - \nu_B$ in spectra of this type owing to the near equality of |L| and |N|, and the main effect of the decreasing chemical shift is an intensity redistribution such that the intensity of the entire spectrum is peaked towards the centre. The difficulties in the assignment problem that occurs in spectra of this type usually arise from inadequate resolution, rendering the number of resolved lines insufficient for an unambiguous assignment. In particular, when $|N| \simeq |L|$, one will seldom be able to resolve line *a* from line *k* or line *b* from line *l*.

With spin-coupling parameters characterizing this type of compound (cf. Eq. (146)), transitions of the two half-spectra are not likely to cross until the spectrum has become so strongly coupled as to approach the A_4 limit, yielding only one broad band at $(\nu_A - \nu_B)/2$.

Schematic AA′ half-spectra of an unsymmetrically 1,2-disubstituted ethane (I) are reproduced in Fig. 56a–e. The spin-coupling parameters assumed in the

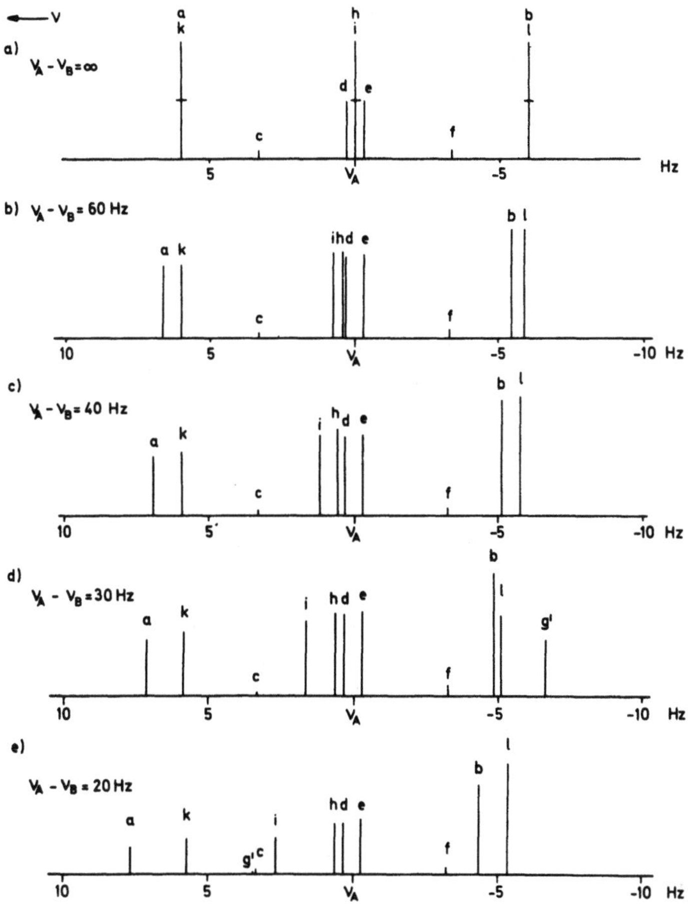

Fig. 56. Low-field half of schematic spectra from a hypothetical unsymmetrical 1,2-disubstituted ethane. The spectra have been calculated with the parameters K = −25 Hz, L = 2 Hz, M = 3 Hz and N = 12 Hz. As in the two previous figures, ν_A is the reference frequency

calculations were:

$$\begin{aligned} K &= -25\,\text{Hz} \\ L &= 2\,\text{Hz} \\ M &= 3\,\text{Hz} \\ N &= 12\,\text{Hz}. \end{aligned} \tag{147}$$

The AA'XX' spectrum of Fig. 56a is almost deceptively simple, since *d*, *e*, *i* and *h* almost coincide at ν_A and transitions *c* and *f* are very weak. Lines *g* and *j* are even weaker and have been left out of the diagram. (They would occur at ±25 Hz relative to ν_A.) A poorly resolved half-spectrum would thus appear as a 1 : 2 : 1 triplet. As the chemical shift decreases to the large but finite value of $\nu_A - \nu_B = 60$ Hz (Fig. 56b), the four antisymmetric transitions *c* through *f* remain completely unchanged. Lines *k* and *l* do not shift away from their AA'XX' positions, but lines *a*, *b* and *h*, *i* undergo a displacement <1 Hz. In fact, the six observed symmetric transitions are indistinguishably from those of an A_2B_2 spin system corresponding to $J_{AB}/(\nu_A-\nu_B)=N/2|\nu_A-\nu_B|=0.1$. This may be rationalized in terms of the energy-level diagram: Owing to the large value of |K|, the $3s_0$ level does not mix with other s_0 levels and the symmetric part of the AA′BB′ level diagram becomes indistinguishable from the 9-level subdiagram [1,1] of an A_2B_2 system. The negligible mixing of the $3s_0$ level with the other s_0 levels also accounts for the negligible intensity of lines *g* and *j*.

As the chemical shift is allowed to decrease even more, the spectrum gradually changes in appearance. This will be particularly noticeable when line *g* approaches line *l′* of the other half-spectrum. It will be seen from the energy-level and transition diagram of Fig. 53 that this will happen when the energy of the $3s_0$ level approaches that of the $2s_0$ level, so that the mixing of the two corresponding states can no longer be neglected. It may be shown that, in order for this mixing to be of importance, the splitting between lines *g* and *l′* must not exceed |L| by an order of magnitude. When this happens, lines *g* and *g′* assume a non-negligible intensity and currently the position of lines *l* and *l′* will be affected, so that the splitting between lines *b* and *l* depends on the degree of mixing. It is in this region that an accurate determination of |K| can easily be made. This is illustrated in Fig. 56d, with $\nu_A-\nu_B=30$ Hz, where *g′* has assumed an intensity comparable to the other lines in the spectrum and should be easily observable.

At further reduction of the relative chemical shift (Fig. 56e, $\nu_A-\nu_B=20$ Hz), the *g*, *g′* lines move further away from the center of the spectrum, gradually declining in intensity until they again become unobservable.

In 1,2-disubstituted ethanes the two nuclei of the symmetrically equivalent pair are attached to the same carbon. In spin systems of type (II), the two symmetrically equivalent nuclei are attached to vicinal carbon atoms and the spectrum assumes an entirely different character. The spectra reproduced in Figs. 57a–c were calculated with the parameters

$$K=16\ \text{Hz},$$
$$L=12\ \text{Hz},$$
$$M=\ 1\ \text{Hz},$$
$$N=-2\ \text{Hz}.$$

In the upper spectrum (Fig. 57a) $\nu_A-\nu_B$ is assumed to be very large, and in the middle spectrum (Fig. 57b) the shift $\nu_A-\nu_B$ is set equal to 60 Hz ($\nu_B=-60$ Hz), while in the lower spectrum (Fig. 57c) $\nu_A-\nu_B=30$ Hz. In the AA′XX′ limit, the two strong lines assigned to transitions *b*, *l* and *a*, *k* respectively will occur close to ν_A owing to the small magnitude of |N|. This is in contrast to the AA′ half-spectra of the three types of spin systems considered above. Furthermore, since $|L|\gg|N|$, the *ab* subspectrum of the four antisymmetric transitions are almost all of the same intensity as in the *ax* limit. On the other hand, the *ab* sub-spectrum of the four transitions *g* through *j* is quite strongly coupled, lines *g* and *j* being the two outermost lines of the entire half-spectrum. As the chemical shift decreases from infinity to the finite value of 60 Hz, the frequencies

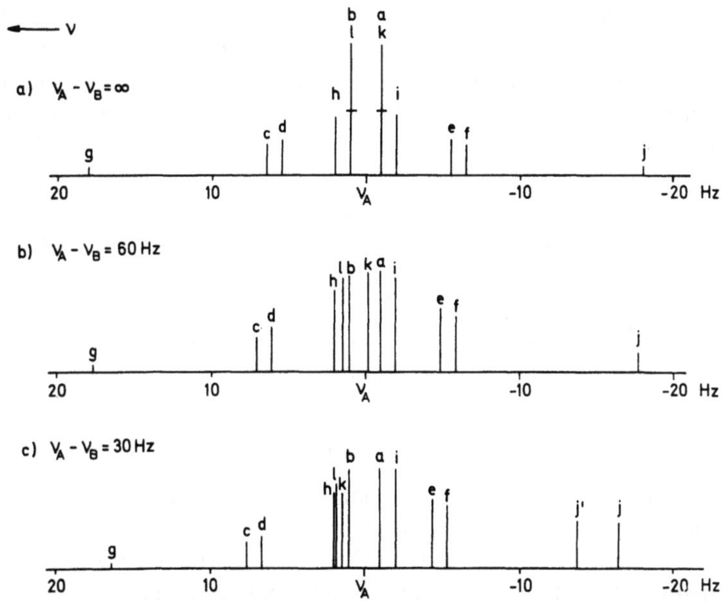

Fig. 57. Low-field half of schematic spectra from a hypothetical symmetrical non-planar ring system, such as that of Fig. (II) in text. The spectra have been calculated with the parameters K = 16 Hz, L = 12 Hz, M = 1 Hz and N = −2 Hz. As in the three previous figures, ν_A is the reference frequency. In diagram c) one line from the high-field half (line j) has been included

of transitions *a*, *b*, and *g* through *j* remain almost unaffected owing to the small value of |M|, but the four antisymmetric transitions *c* through *f* and transitions *k* and *l* are shifted away from $(\nu_A + \nu_B)/2$ by ≈0.6 Hz. Because |N| is so small, the two transitions *b* and *a* will remain in the vicinity of ν_A and have nearly equal intensities, even for quite small values of $\nu_A - \nu_B$. Similarly, lines *i* and *h* will move very little with respect to ν_A as ν_B approaches ν_A, so that lines *h*, *b*, *a*, and *i* are in approximately the same configuration for every chemical shift owing to the small value of |M| (Fig. 57c). The four antisymmetric transitions *c* through *f* will behave approximately as in a spectrum for which |M| = 0. Thus lines *c* and *e* together with *c'* and *e'* will yield an *ab* type subspectrum with effective shift and coupling parameters equal to $|\nu_A - \nu_B|$ and |L| respectively. Lines *d*, *f*, *d'*, and *f'* yield a similar *ab* type subspectrum displaced from the former by |M| approximately.

The behaviour of lines *g*, *h*, and *j* is best understood with reference to the transition and energy level diagram. As the chemical shift decreases, the frequency of line *j* will approach that of line *k'* (and similarly *j'* will approach *k*). This will entail strong mixing of levels $3s_0$ and $1s_0$. The mixing will result in an increase of the intensities of lines *j* and *j'* at the expense of lines *k* and *k'*. As compared with their position in the AA'XX' limits, lines *g* and *j* will move inwards towards ν_A. The magnitude of these displacements will be approximately the same and line *k* will move outwards to the same extent.

These effects on the transitions g, j, and k can be seen in Fig. 57c which has been calculated with $\nu_A - \nu_B = 30$ Hz. As the chemical shift becomes small enough, line k tends to become the outermost line in the spectrum and drop to a very small intensity, lines j and j' will cross, and finally line l will move slightly outwards as the chemical shift decreases but will never receed very far from ν_A. All the effects discussed in this last paragraph occur over a fairly wide range of chemical shifts $|\nu_A - \nu_B|$ owing to the large magnitude of $|L|$. The mixing between levels $1s_0$ and $3s_0$ will be at its largest when the chemical shift is such as to make the expression

$$\nu_A - \nu_B + N/2 - K$$

vanish, as may be deduced from the AA′XX′ expressions for the frequencies of lines j and k' (cf. Table 11 in Section V.h).

Example 1: The AA′BB′ Proton Resonance Spectrum of 1,2-dibromobenzene

The complete 60 MHz proton resonance spectrum of the title compound in CS_2 solution (26% by weight) is shown in Fig. 58. The compound belongs to category IV of Section IV.9.f, the typical spectra of which are shown in Fig. 54. A comparison between the experimental spectrum and those of Fig. 54b and c suggests the assignment of labels given in Fig. 58. This assignment may now be checked by means of Eqs. (212) and (124). Using the line positions in Fig. 58, we obtain $c + f = 27.03$ Hz and $d + e = 26.97$ Hz. These sums agree within the estimated uncertainty in the line positions (± 0.08 Hz) – this uncertainty being largest for the unresolved lines d and f.

From Eq. (124) we also obtain $a + b = 27.79$ Hz, $g + j = 27.78$ Hz and $i + h = 27.77$ Hz. Our assignment in Fig. 58 thus seems to be correct. As a further check, we may use the intensity relations in Eqs. (125) to (128). From Eq. (128) we obtain $I_a/I_b = 0.48$ which compares favourably with the experimental value of 0.51. Eqs. (125) and (126) imply that $I_a + I_b = 2(I_c + I_e)$. The experimental ratio of these intensity sums is 2.3. The remaining Eqs. (126) and (127) cannot be used in the present case due to overlap of lines.

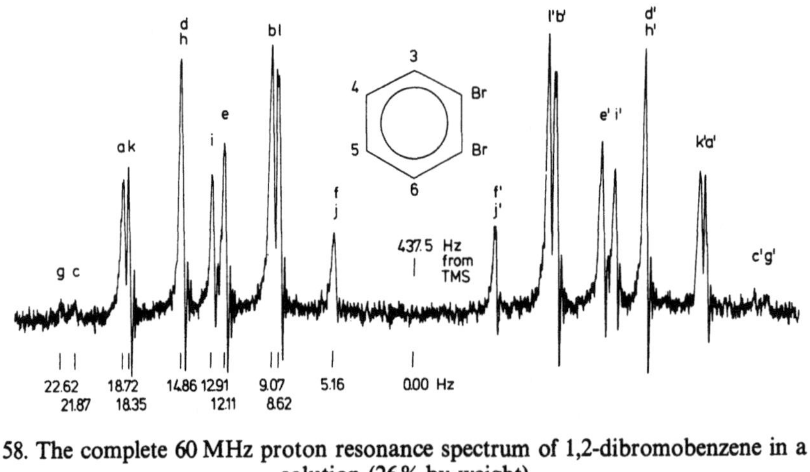

Fig. 58. The complete 60 MHz proton resonance spectrum of 1,2-dibromobenzene in a CS_2 solution (26% by weight)

The NMR parameters may now be evaluated from Eqs. (136) to (141). The labelling of transitions in Fig. 58 conforms with the inequalities $c>d>e>f$ and $a>b$ of Eqs. (129) and (130), which leads to *positive* signs of parameters L, M and N. The average parameter values are

$$|\nu_A - \nu_B| = 26.05 \text{ Hz},$$
$$K = 7.83 \text{ Hz},$$
$$L = 6.54 \text{ Hz},$$
$$M = 7.21 \text{ Hz},$$
$$N = 9.60 \text{ Hz},$$

which gives the following spin-coupling constants

$$J_{AB} = 8.07 \text{ Hz},$$
$$J_{A'B} = 1.53 \text{ Hz},$$
$$J_{AA'} = 7.52 \text{ Hz},$$
$$J_{BB'} = 0.31 \text{ Hz}.$$

These parameters can be assigned to the following couplings in the molecule

$$J_{AB} = J_{34} = J_{56},$$
$$J_{A'B} = J_{35} = J_{46},$$
$$J_{AA'} = J_{45},$$
$$J_{BB'} = J_{36}.$$

The analysis gives only the absolute value of the relative chemical shift $\nu_A - \nu_B$, and it cannot be determined whether the protons adjacent to the bromine atoms give rise to the low-field or high-field half of the AA'BB' spectrum. Comparison with other data, however, indicates that the 3 and 6 protons are the less shielded pair and give rise to the low-field part of the experimental spectrum. Thus, $\nu_A = \nu_4 = \nu_5 = 424.5$ Hz and $\nu_B = \nu_3 = \nu_6 = 450.6$ Hz from TMS respectively.

Example 2: The AA'BB' Proton Resonance Spectrum of 4-bromonitrobenzene

The complete 56.4 MHz proton resonance spectrum of the title compound $CDCl_3$ solution (6% by weight) is shown in Fig. 59. The compound belongs to category III of Section IV.9.f and typical spectra of this class of compounds are shown in Fig. 55. The AA'BB' spectrum should be symmetrical; however, the two halves of the experimental spectrum are not complete mirror images of each other, more structure has been resolved in the high-field half. The reason for this is presumably a spin coupling of the nitrogen nucleus to the H-2 and H-6 protons. The relaxation rate of the ^{14}N nucleus is not sufficiently rapid to completely extinguish the effects of this coupling (cf. the spectrum of formamide in Fig. 9). The asignment should in this case be based on the better resolved half of the spectrum.

A comparison between the experimental spectrum and those of Fig. 55 allows a direct assignment of the resolved transitions *d*, *e* and *g* through *j*. The intense signal about −17 Hz from $(\nu_A + \nu_B)/2$ shows a pronounced shoulder on the low-field side. From a consideration of the intensities in Fig. 55, it is most likely that this shoulder should be attributed to transition *c*, while *a* and *k* overlap and give rise to the strongest peak in this region. The doublet observed about −8 Hz is due to transitions *b*, *f*, and *l*, and a comparison of the experimental intensities with those of Fig. 55 shows that transition *f* coincides with either *b* or *l* at −7.94 Hz. A distinction between these two assignments can be made by use of the frequency sum rules in Eqs. (123) and (124).

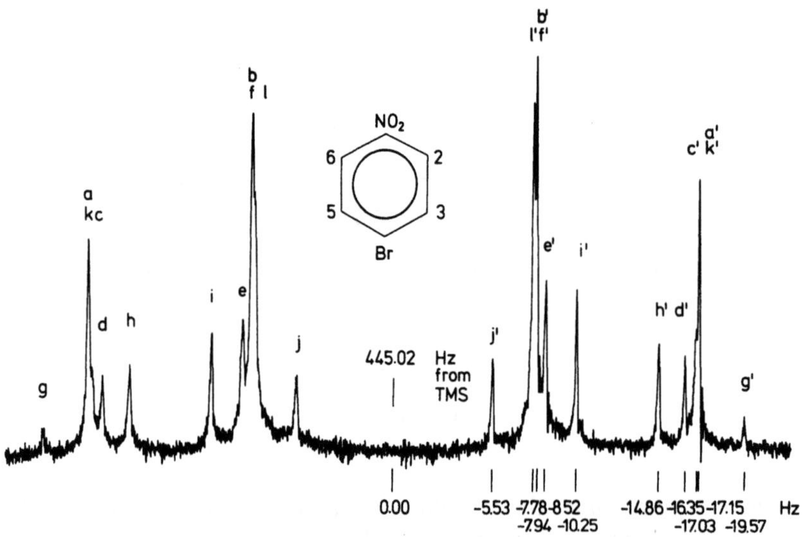

Fig. 59. The complete 56.4 MHz proton resonance spectrum of 4-bromonitrobenzene in a $CDCl_3$ solution (6% by weight)

Using the line positions in Fig. 59, we obtain $g+j=25.10$ Hz and $i+h=25.11$ Hz. The frequency sum $a+b$ can now be evaluated either as $17.15+7.78=24.93$ Hz or as $17.15+7.94=25.09$ Hz, and from the frequency sum rule of Eq. (124) it is clear that the latter assignment with $b=7.94$ Hz is the correct one. From this we may conclude that $l=7.78$ Hz and that *f* overlaps the transition *b* at 7.94. This may now be checked by means of the frequency sum rule of Eq. (121) which gives $c+f=24.97$ Hz and $d+e=24.87$ Hz. In view of the fact that both lines *c* and *f* overlap other lines, the small discrepancy in the frequency sum above is well within reasonably estimates of errors in line positions. The final assignment is given in Fig. 59.

In the present example, the intensity sum rules of Eqs. (125) to (127) are of limited value due to overlap of lines.

The NMR parameter may now be evaluated from Eqs. (136) to (141). The labelling of transitions in Fig. 59 conforms with the inequalities of Eqs. (129) and (130), giving positive signs for parameters L, M, and N. The average parameter value are

$$|\nu_A - \nu_B| = 23.36\ \text{Hz},$$
$$K = 4.88\ \text{Hz},$$
$$L = 8.43\ \text{Hz},$$
$$M = 0.65\ \text{Hz},$$
$$N = 9.60\ \text{Hz},$$

which gives the following spin couplings

$$J_{AB} = 8.82\ \text{Hz},$$
$$J_{A'B} = 0.39\ \text{Hz},$$
$$J_{AA'} = 2.77\ \text{Hz},$$
$$J_{BB'} = 2.12\ \text{Hz}.$$

We have already assigned the low-field half of the spectrum to the protons *ortho* to the nitrosubstituent. This is also in agreement with the well-known deshielding effect of this group on the hydrogens in aromatic compounds. From the analysis we cannot differentiate between the two *meta* couplings $J_{AA'}$ and $J_{BB'}$, however, we assign the smaller of the two to the coupling across the electron-withdrawing nitro group. This gives the assignment

$$J_{AB} = J_{23} = J_{56},$$
$$J_{A'B} = J_{25} = J_{36},$$
$$J_{AA'} = J_{35},$$
$$J_{BB'} = J_{26},$$

and the chemical shifts

$$\nu_A = \nu_3 = \nu_5 = 433.34\ \text{Hz}$$

and

$$\nu_B = \nu_2 = \nu_6 = 456.70\ \text{Hz}$$

from TMS respectively.

Example 3: The AA′BB′ Proton Resonance Spectrum of Ethylene Sulphite

The low-field half of the 100 MHz proton resonance spectrum of ethylene sulphite (V) (ca. 15% by weight in $CDCl_3$) is shown in Fig. 60.

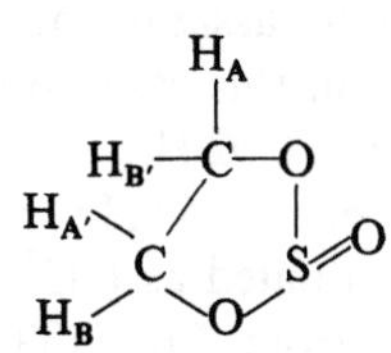

(V)

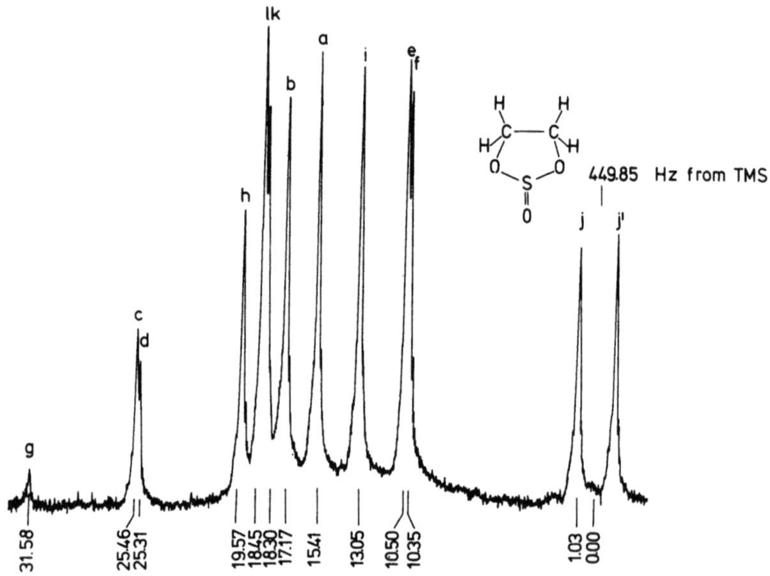

Fig. 60. The low-field half of the 100 MHz proton resonance spectrum of ethylene sulphite in a $CDCl_3$ solution (ca. 15% by weight)

The compound belongs to category II of Section IV.9.f and typical spectra of this class of compounds are shown in Fig. 57. A comparison between the experimental spectrum and the spectra of Fig. 57 suggests the assignments of lines c through f and g, j as given in Fig. 60. This assignment agrees very well with the frequency sum rule of Eq. (121): $c+f=d+e=35.81$ Hz. This gives an unambiguous value of $g+j=32.61$ Hz to be used to check tentative assignments of transitions a, b, i, and h through the sum rules of Eq. (124). This frequency sum rule is consistent with two assignments of transitions a, b, i, and h. Either $a=15.41$ Hz, $b=17.17$ Hz, $i=13.05$ Hz, and $h=19.57$ Hz, *or* $a=13.05$ Hz, $b=19.57$ Hz, $i=15.41$ Hz, and $h=17.17$ Hz, and a differentiation between these two assignments cannot be made at this stage.

Evaluation of the chemical shift $\nu_A-\nu_B$ from $\sqrt{2(ce+df)}$ (Eq. (136)) yields $\nu_A-\nu_B=32.53$ Hz. Evaluation of $\nu_A-\nu_B$ as $2\sqrt{ab}$ of Eq. (136) gives $\nu_A-\nu_B=32.53$ Hz using the assignment $a=15.41$ Hz, $b=17.17$ Hz, or $\nu_A-\nu_B=32.68$ Hz using $a=13.05$ Hz, $b=19.57$ Hz. This suggests that the former assignment is the correct one. In the present case, where all lines are well resolved and their frequencies accurately measured, the discrepancy between the chemical shifts calculated from the two assignments is large enough to allow us with some confidence to disregard the second assignment. This is further confirmed by evaluating N from Eq. (140) since in this case no ambiguity persists in the value of $C=a+b=i+h=g+j=32.60\pm0.02$ Hz. $N=a-b$ in the two discussed assignments becomes -1.76 Hz or -6.52 Hz, respectively; of these, only the first value is consistent with N evaluated as $\sqrt{C^2-(\nu_A-\nu_B)^2}$.

$|M|$ may now be calculated from Eq. (139), yielding 0.17 Hz, and from Eq. (138) we obtain $|L|=14.96$ Hz.

In order to calculate the value of K, we must make an assignment of transitions k and l to the two remaining lines in the spectrum. Some guidance may be obtained from the theoretical spectrum in Fig. 56c, where it is seen that at $\nu_A - \nu_B = 30$ Hz line l is stronger than line k. It may also be noted that with decreasing chemical shift line k moves away from $(\nu_A + \nu_B)/2$ and has at $\nu_A - \nu_B = 30$ Hz almost reached line i. We cannot, however, altogether rely on the observed intensity of overlapping lines, since field inhomogeneity and finite sweep rates may introduce distortion. As our first set we take $k = 18.30$ Hz and $l = 18.45$ Hz, and as our second $k = 18.45$ Hz and $l = 18.30$ Hz.

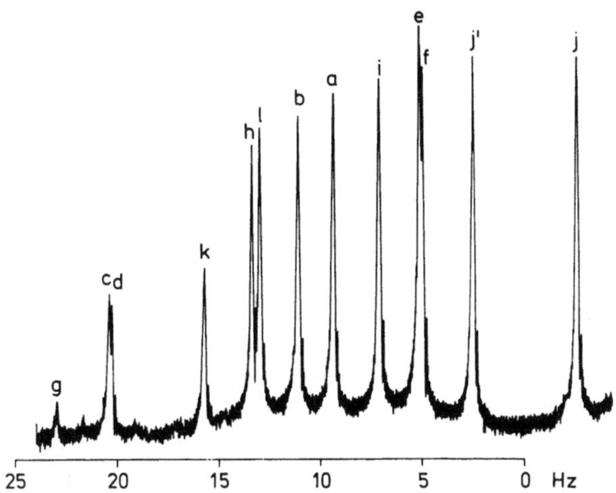

Fig. 61. The 60 MHz spectrum of ethylene sulphite

With the first set we obtain from Eq. (137) $k = 13.62$ Hz and with the second set we get $k = 13.92$ Hz. In order to distinguish between these two sets, a spectrum was also recorded at 60 MHz, where the lower relative chemical shift $\nu_A - \nu_B$ should make possible a distinction of transitions k and l. This spectrum is shown in Fig. 61. The gradual displacement of transition j towards the BB' half of the AA'BB' spectrum with decreasing $\nu_A - \nu_B$, apparent in Fig. 57, will ultimately result in a crossing over of j and its counterpart j' at a certain relative chemical shift $\nu_A - \nu_B$. This has evidently happened in Fig. 61, since the sum rule of Eq. (124) can otherwise never be fulfilled. When this has been realized, the transitions a through j can be assigned in a straightforward manner employing the sum rules of Eqs. (121) and (124) and comparison with the assignment in Fig. 60. The two remaining transitions can now be assigned in a unique way, since the value of k evaluated from Eq. (137) will differ by about 5.5 Hz between the assignment indicated in Fig. 61 and that when the labels k and l are reversed. The value of k evaluated from the 60 MHz spectrum is 13.65 Hz, in good agreement with the value 13.62 Hz obtained with our first set in the case of the 100 MHz spectrum. This assignment is also indicated in Fig. 60. It should be noted that the assignment given in Figs. 60 and 61 conforms with the conventions of Eqs. (129) and (131), *i.e.* K, L, and M come out positive while N may take either sign; in the present case, N is negative.

To summarize, the parameters obtained are

$$|\nu_A - \nu_B| = 32.53\ \text{Hz},$$
$$K = 13.62\ \text{Hz},$$
$$L = 14.96\ \text{Hz},$$
$$M = 0.17\ \text{Hz},$$
$$N = -1.76\ \text{Hz},$$

which gives the following spin couplings

$$J_{AA'} = 6.90\ \text{Hz},$$
$$J_{BB'} = 6.73\ \text{Hz},$$
$$J_{AB} = 6.60\ \text{Hz},$$
$$J_{AB'} = -8.36\ \text{Hz}.$$

As pointed out in Section IV.9.a, *the signs of L and M depend only on the labelling of the spins and have no physical significance.* Explicitly this means that one may *not* conclude from the set of coupling constants given above that the geminal coupling constant is $J_{AB} = +6.60$ Hz nor -6.60 Hz. The geminal coupling could equally well be $J_{AB'} = -8.36$ Hz or even $+8.36$ Hz (if the signs of the couplings are all reversed). Similarly, one cannot say which one of the two vicinal couplings $J_{AA'}$ and $J_{BB'}$ correspond to the protons at lower fields. As always in AA′BB′ spectra, one has to have recourse to other information, such as observed regularities of NMR parameters, isotopic labelling etc. In the present case such information may be obtained from the NMR spectrum of (VI) which has been studied by PRITCHARD and LAUTERBUR (J. Am. Chem. Soc. **83**, 2105 (1961)). They observed a geminal coupling of ± 8.5 Hz and assigned the low-field proton to the proton *cis* to the $S=O$ group. This geminal coupling agrees well with our coupling $J_{AB'}$. The negative sign of this geminal coupling is in accordance with the sign inferred for geminal couplings in analogue spin systems (cf. L. M. JACKMAN and S. STERNHELL, "Nuclear Magnetic Resonance Spectroscopy in Organic Chemistry", Pergamon Press, London, 1969) as are also the positive signs of the vicinal couplings $J_{AA'}$, $J_{BB'}$, and J_{AB}. No information is available at present which would allow an assignment of the vicinal couplings $J_{AA'}$ and $J_{BB'}$. The labelling of the protons as AA′ and BB′ indicated in (V) above is therefore tentative, and the labels of A and B may actually be the opposite.

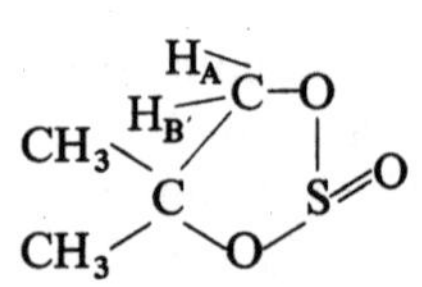

(VI)

Example 4: The AA′BB′ Proton Resonance Spectrum of 2-bromoethanol

The low-field half of the 56.4 MHz proton resonance spectrum of 2-bromoethanol in a $CDCl_3$ solution (15% by weight) is shown in Fig. 62. The compound belongs to category I of Section IV.9.f, and typical spectra of this class of

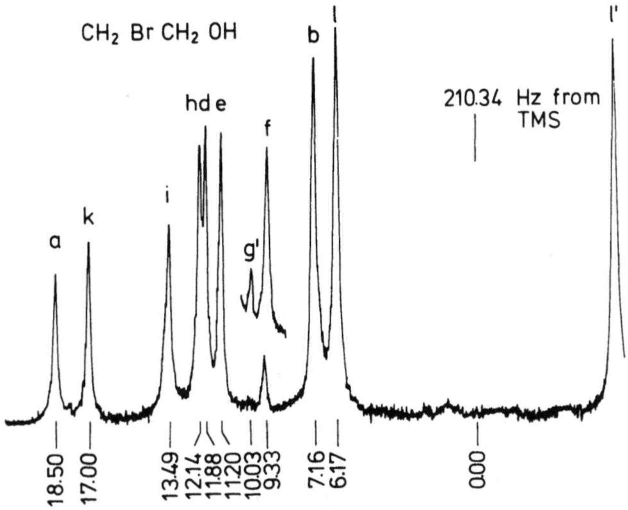

Fig. 62. The low-field half of the 56.4 MHz proton resonance spectrum of 2-bromoethanol in a $CDCl_3$ solution (15% by weight)

compounds are shown in Fig. 56. A comparison of the experimental spectrum with those of Fig. 56 suggests the assignment of lines *a*, *b*, *d*, *e*, *h*, *i*, *k*, and *l* as given in Fig. 62. This assignment gives $a+b=25.66$ Hz and $i+h=25.63$ Hz, which agrees well with the frequency sum rule of Eq. (124). The frequency sum of Eq. (123) gives $d+e=23.08$ Hz, which should be checked against the assignment of lines *c* and *f*. By recording the spectrum at a higher rf field, two lines can be observed in the 9–10 Hz region (inserted in Fig. 62) of which one should be the *f* transition. By putting $f=9.33$ Hz, we obtain from Eq. (123) $c=13.75$ Hz, *i.e.* the *c* line should fall close to line *i*, where it may indeed be observed as a shoulder on the low-field side. From the theoretical spectra in Fig. 56, it is seen that the position of line g' is highly dependent on the relative chemical shift $\nu_A-\nu_B$. In Fig. 56d the g' transition is strong and has just entered the low-field half of the spectrum. At further reduction of the chemical shift (Fig. 56e), line g' becomes weaker and moves further down-field. A comparison with the experimental spectrum in Fig. 62 indicates that the weak line at 10.03 Hz should be assigned as the g' transition, *i.e.* our experimental spectrum is recorded at a relative chemical shift somewhere in the region between Fig. 56d and 56e.

The resolution of the g' transition is necessary to render possible an evaluation of the parameter K (cf. Eq. (137)). At a large relative chemical shift $\nu_A-\nu_B$ the g' transition becomes weak and unobservable. This occurs when the spectrum of the present compound is recorded at 100 MHz.

Evaluating the NMR parameters from Eqs. (136) to (141), we obtain

$$|\nu_A-\nu_B| = 23.02\ \text{Hz},$$

$$K = -22.68\ \text{Hz},$$

$$L = 1.73\ \text{Hz},$$

$$M = 1.87\ \text{Hz},$$

$$N = 11.34\ \text{Hz}.$$

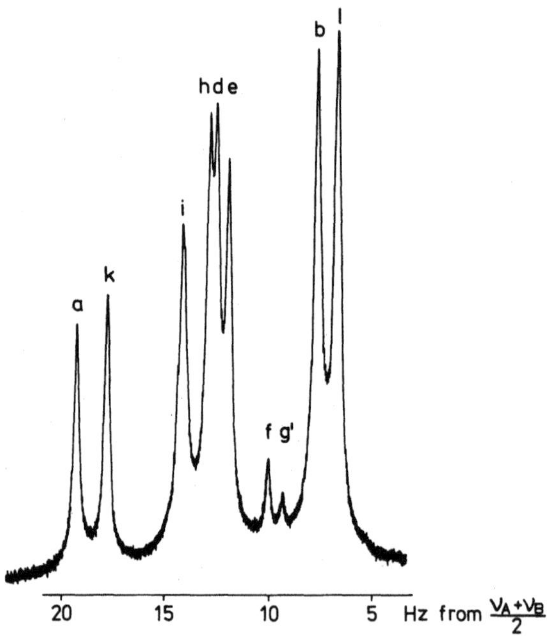

Fig. 63. The 60 MHz spectrum of 2-bromoethanol

To check the correctness of the assignment of the f and g' transitions, a spectrum was recorded also at 60 MHz (Fig. 63). This gives only a small increase in the relative chemical shift $\nu_A - \nu_B$, and the two spectra of Figs. 62 and 63 are indeed very similar. An exception is the region around 10 Hz, where it can be seen that the g' transition in Fig. 63 has shifted to the high-field side of the f line and at the same time multiplied in intensity as compared to the spectrum in Fig. 62. This amply illustrates how sensitive the g' transition is to the relative chemical shift, and that it is only within a fairly narrow $\nu_A - \nu_B$ region that a determination of K can be made in this type of compound.

The spin-coupling constants obtained are

$$J_{AB} = 6.54 \text{ Hz},$$
$$J_{AB}' = 4.81 \text{ Hz},$$
$$J_{AA'} = -10.41 \text{ Hz},$$
$$J_{BB'} = -12.28 \text{ Hz}.$$

On the basis of the relative chemical shifts in similar bromides and alcohols, the low field multiplet can be assigned to the CH_2OH methylenes. The geminal coupling constants in CH_2 groups attached to oxygen are generally found to be more negative than the corresponding coupling constant in CH_2Br fragments (cf. A. A. BOTHNER-BY in "Advances in Magnetic Resonance", Vol. I (J. S. WAUGH, Editor) Academic Press, New York, 1965). We therefore tentatively assign $J_{AA'}$ to the CH_2Br grouping and $J_{BB'}$ to the CH_2OH grouping.

V. Quantum Mechanical Formalism

V.1. Introduction to the Quantum Mechanical Description of Spin Systems

The basic quantum mechanics required will be described with a sequence of postulates and definitions with a few comments interpolated. The treatment does not claim maximum rigour and elegance but is intended to be useful for the practical analysis of NMR spectra.

Postulate 1. The state of the spin system is represented mathematically by a wave function ψ. The function $c\psi$, where c is an arbitrary constant, represents the same state as ψ itself.

Comments. It is not necessary for us to specify the variables on which ψ depends; in fact, this is never done for spin wave functions.

Definition 1. A number of functions, $\psi_1, \psi_2, \psi_3 \dots \psi_n$, are said to be linearly independent if the identity

$$\sum_{j=1}^{n} c_j \psi_j \equiv 0$$

cannot be fulfilled unless all the coefficients c_j vanish. Otherwise the functions are said to be linearly dependent, and in such case it is possible to express at least one of the functions as a linear combination of the remaining ones.

Definition 2. A set of functions ψ_j will be said to form a *complete basis* for the states of the spin system if any admissible state of the spin system can be written as a linear combination of the basis functions.

Postulate 2. The spin states of a particle of spin 1/2 are always expressible by a linear combination of two basis functions. These may be chosen as the states corresponding to spin up and spin down. These two functions will henceforth be called α and β. By forming two linearly independent combinations of α and β, for example $\psi_1 = \alpha \sin\varphi + \beta \cos\varphi$ and $\psi_2 = \alpha \cos\varphi - \beta \sin\varphi$, where φ is an arbitrary angle, we may form a new pair of basis functions.

Definition 3. For any given pair of spin functions ψ_j and ψ_k we may define a complex number called the *inner product* and designated by the symbol

$$\langle \psi_j | \psi_k \rangle .$$

The order of the functions is important because $\langle \psi_k | \psi_j \rangle$ is defined as the complex conjugate of $\langle \psi_j | \psi_k \rangle$. The inner product is linear in the latter function so that, if $\psi_k = c\psi_m + d\psi_n$, then the inner product $\langle \psi_j | \psi_k \rangle$ is given by

$$\begin{aligned} \langle \psi_j | \psi_k \rangle &= \langle \psi_j | c\psi_m + d\psi_n \rangle \\ &= c\langle \psi_j | \psi_m \rangle + d\langle \psi_j | \psi_n \rangle . \end{aligned}$$

Definition 4. Two functions are said to be orthogonal if their inner product vanishes.

Comment. The orthogonality of two wave functions may be interpreted to mean that none of the properties of one of the states is contained in the other and *vice versa.*

Definition 5. The *norm* of a spin function is defined as the inner product of the spin function with itself, *i.e.*

$$\langle \psi_j | \psi_j \rangle .$$

The function will be said to be *normalized* if the norm equals unity. This can always be arranged by a judicious choice of constant c in Postulate 1.

Postulate 3. The basis functions α and β are normalized and mutually orthogonal *i.e.*

$$\langle \alpha | \alpha \rangle = \langle \beta | \beta \rangle = 1 ,$$
$$\langle \alpha | \beta \rangle = \langle \beta | \alpha \rangle = 0 .$$

For a system containing many spins, a complete set of basic functions is obtained by forming all linearly independent products of the basis functions of the individual spins. Consider, for example, a spin system consisting of two spins, labelled A and B. The basis functions for spin A may be written $\alpha(\mathrm{A})$ and $\beta(\mathrm{A})$ and those of spin B may be written $\alpha(\mathrm{B})$ and $\beta(\mathrm{B})$. The basis of the two-spin system consists of the four functions

$$\alpha(\mathrm{A})\,\alpha(\mathrm{B}) ,$$
$$\alpha(\mathrm{A})\,\beta(\mathrm{B}) ,$$
$$\beta(\mathrm{A})\,\alpha(\mathrm{B}) ,$$
$$\beta(\mathrm{A})\,\beta(\mathrm{B}) .$$

These four functions are all normalized and pairwise orthogonal. Usually, the index numbers in brackets are suppressed and the order of occurrence of the basis functions in the product is used as a label of the corresponding spin.

Definition 6. A linear operator Q is a prescription according to which every spin function ψ_j is mapped into a new function ψ'_j, or symbolically

$$\psi'_j = \mathrm{Q}\psi_j .$$

The linearity of the operation implies that linear relations between functions remain invariant under the mapping, thus

$$\mathrm{Q}(c\psi_j + d\psi_k) = c\mathrm{Q}\psi_j + d\mathrm{Q}\psi_k = c\psi'_j + d\psi'_k .$$

Comment. To know the effect of a *linear* operator on any spin function it is sufficient to known the effect of the operator on every *basis* function.

Definition 7. The effect of a sum of operators operating on a function ψ_j is the sum of the functions produced by the individual operators.

Definition 8. A product of operators operating on a function ψ_j is defined as successive operations by the operators starting with the one nearest the function;

thus, $PQ\psi_j = \psi_j''$ where $\psi_j'' = P\psi_j'$ and where $\psi_j' = Q\psi_j$. If $PQ\psi_j$ is the same function as $QP\psi_j$ for any ψ_j, then the operators P and Q are said to commute. More generally one may define an operator known as the *commutator* of P and Q

$$[P, Q] = PQ - QP.$$

When a spin system is subject to any kind of interaction, the state of the spin system usually changes and hence it is natural that interactions should be represented mathematically as operators. In quantum mechanics these operators are always linear.

Postulate 4. Interactions and observable entities are represented in quantum mechanics by linear operators.

To represent an observable it is not sufficient for an operator to be linear, but it must also be *Hermitian* according to Definition 9.

Definition 9. An operator Q is said to be *Hermitian* if $\langle\psi_j|Q\psi_k\rangle$ is the complex conjugate of $\langle\psi_k|Q\psi_j\rangle$. If an operator Q is not Hermitian, one may define a *Hermitian adjoint* operator Q^+ such that $\langle\psi_k|Q^+\psi_j\rangle$ is the complex conjugate of $\langle\psi_j|Q\psi_k\rangle$. The inner product $\langle\psi_j|Q\psi_k\rangle$ is often written $Q_{jk} = \langle\psi_j|Q|\psi_k\rangle$ and is called the "*jk*"-th matrix element of Q in the basis ψ_j.

Definition 10. If Q is an operator and q_j is a constant, and if, furthermore, ψ_j is a function such as to satisfy the equation

$$Q\psi_j = q_j\psi_j$$

then ψ_j is said to be an *eigenfunction* belonging to the *eigenvalue* q_j of the operator Q. The eigenvalues of a Hermitian operator are always real numbers and eigenfunctions belonging to different eigenvalues are orthogonal.

Postulate 5. If the wave function ψ is an eigenfunction ψ_j belonging to the eigenvalue q_j of the operator Q, then q_j is the only possible value for the observable Q. If, on the other hand, the state of the spin system is *not* an eigenstate of the observable Q, then the observed values of Q may be any of its eigenvalues with a probability $|c_j|^2$ for the value q_j, where c_j is the coefficient of the eigenfunction ψ_j in the expansion of the actual wave function ψ along the basis formed by the eigenfunctions of Q.

V.2. Quantum Mechanical Calculation of NMR Spectra

a) The NMR Spin Hamiltonian

The most important observable in NMR spectroscopy is the energy of the spin system, and the corresponding operator is called the NMR spin Hamiltonian, $\mathscr{H}$. It takes the form

$$\mathscr{H} = \sum_i \nu_i I_z(i) + \sum_{i<j} J_{ij}\boldsymbol{I}(i)\cdot\boldsymbol{I}(j). \qquad (1)$$

In Eq. (1) both members have been divided by Planck's constant *h*, so that the constant parameters ν_i and J_{ij} have the dimension of frequency. The significance of these parameters has been explained earlier, ν_i being the Larmor frequency of nucleus i and J_{ij} being the spin-coupling constant between nuclei i and j. The vector operators $\boldsymbol{I}(i)$ and $\boldsymbol{I}(j)$ are the nuclear spin angular momentum operators with components I_x, I_y, and I_z, and the scalar product $\boldsymbol{I}(i) \cdot \boldsymbol{I}(j)$ may be expanded according to

$$\boldsymbol{I}(i) \cdot \boldsymbol{I}(i) = I_x(i)\, I_x(j) + I_y(i)\, I_y(j) + I_z(i)\, I_z(j)\,. \qquad (2)$$

Every spin angular momentum operator operates solely on the spin function belonging to the same spin. In terms of the basis functions α and β their effects are summarized in Table 6.

Table 6. *Effects of angular momentum operators on simple spin functions*

Operator Q	Spin functions α	β
I_x	$\beta/2$	$\alpha/2$
I_y	$i\beta/2$	$-i\alpha/2$
I_z	$\alpha/2$	$-\beta/2$

It is seen from Table 6 that the spin functions α and β are eigenfunctions of I_z, the z component of angular momentum, where the z direction is defined as the direction of the external magnetic field H_0. Instead of the transverse components I_x and I_y, one customarily introduces non-Hermitian linear combinations of these operators, $I_+ = I_x + i \cdot I_y$ and $I_- = I_x - i \cdot I_y$ where $i = \sqrt{-1}$. These operators are known as ladder operators because they have the following properties.

$$\begin{aligned} I_+\alpha &= 0 \\ I_+\beta &= \alpha \\ I_-\alpha &= \beta \\ I_-\beta &= 0\,. \end{aligned} \qquad (3)$$

In terms of these operators the spin Hamiltonian may be rewritten:

$$\mathscr{H} = \sum_i \nu_i I_z(i) + \sum_{i<j} J_{ij}[I_z(i)\, I_z(j) + I_+(i)\, I_-(j)/2 + I_-(i)\, I_+(j)/2]\,. \qquad (4)$$

The possible energy values of the spin system are the eigenvalues of the Hamiltonian operator, and the possible transition frequencies are obtained as the differences between these eigenvalues. The first problem in calculating an NMR spectrum is to solve the eigenvalue problem of the corresponding Hamiltonian. The eigenfunctions correspond to the *stationary states* of the spin system.

b) The Transition Matrix Elements and Relative Intensities

Suppose that one has solved the eigenvalue problem of the spin Hamiltonian

$$\mathscr{H}\psi_j = E_j\psi_j \qquad (5)$$

where ψ_j is the eigenfunction belonging to the eigenvalue E_j of $\mathscr{H}$. The number of linearly dependent eigenfunctions equals the number of basis functions of the corresponding spin system. For example, a two-spin system has four basis functions and thus four linearly independent stationary states. The number of possible transitions thus equals $(4 \times 3)/2 = 6$, but only four transitions are allowed. This is so because the transition probability, I_{jk}, may be shown to be proportional to the square of the matrix element of $F_x = \sum_i I_x(i)$ between the initial and final states

$$I_{jk} = |\langle \psi_j | F_x \psi_k \rangle|^2 . \tag{6}$$

It may be shown that the matrix element vanishes unless the magnetic quantum numbers of j and k differ by unity.

c) An Example: The AB Spin System

A convenient set of basis functions for the two-spin system are the simple product functions

$$\varphi_1 = \alpha\alpha ,$$
$$\varphi_2 = \alpha\beta ,$$
$$\varphi_3 = \beta\alpha ,$$
$$\varphi_4 = \beta\beta .$$

A general wave function for the two-spin system may then be obtained as a linear combination of these four basis functions.

The Hamiltonian of the two-spin system may be written:

$$\begin{aligned} \mathscr{H} = \nu_A I_z(A) + \nu_B I_z(B) + J_{AB} I_z(A)\, I_z(B) \\ + J_{AB}[I_+(A)\, I_-(B) + I_-(A)\, I_+(B)]/2 . \end{aligned} \tag{7}$$

The matrix elements of the Hamiltonian operator are readily obtained by use of the orthonormality of the basis functions and by calculating the results obtained by operating with the Hamiltonian on the basis functions, keeping in mind the rules laid down in Table 6 and in Eq. (3). The results obtained are

$$\mathscr{H}\varphi_1 = (\nu_A/2 + \nu_B/2 + J_{AB}/4)\, \varphi_1 = E_1 \varphi_1 , \tag{8a}$$
$$\mathscr{H}\varphi_2 = (\nu_A/2 - \nu_B/2 - J_{AB}/4)\, \varphi_2 + (J_{AB}/2)\, \varphi_3 , \tag{8b}$$
$$\mathscr{H}\varphi_3 = (-\nu_A/2 + \nu_B/2 - J_{AB}/4)\, \varphi_3 + (J_{AB}/2)\, \varphi_2 , \tag{8c}$$
$$\mathscr{H}\varphi_4 = (-\nu_A/2 - \nu_B/2 + J_{AB}/4)\, \varphi_4 = E_4 \varphi_4 . \tag{8d}$$

From Eq. (8a) through (8d) it is evident that φ_1 and φ_4 are eigenfunctions of $\mathscr{H}$ belonging to the eigenvalues $E_1 = H_{11}$ and $E_4 = H_{44}$ enclosed in brackets in Eqs. (8a) and (8d). On the other hand, the result of operating with $\mathscr{H}$ on φ_2 yields a linear combination of φ_2 and φ_3 which may be written

$$\mathscr{H}\varphi_2 = H_{22}\varphi_2 + H_{23}\varphi_3 \tag{9a}$$

and similarly

$$\varphi_3 = H_{33}\varphi_3 + H_{32}\varphi_2 \tag{9b}$$

so that φ_2 and φ_3 are obviously not eigenfunctions because of the occurrence of the off-diagonal matrix element $H_{23} = H_{32} = J_{AB}/2$. The true eigenfunctions φ_2 and φ_3 must be orthogonal to the eigenfunctions φ_1 and φ_4 and hence can be linear combinations solely of φ_2 and φ_3. Since φ_2 and φ_3 are normalized and orthogonal, one may write any pair of linear combinations that should also remain normalized and orthogonal in terms of the arbitrary angle θ

$$\psi_2 = \varphi_2 \cos\theta + \varphi_3 \sin\theta\,, \tag{10a}$$

$$\psi_3 = -\varphi_2 \sin\theta + \varphi_3 \cos\theta\,. \tag{10b}$$

The two functions ψ_2 and ψ_3 must obey the eigenvalue equation

$$\mathscr{H}\psi = E\psi\,.$$

By virtue of Eqs. (9a–b) this beomes

$$\begin{aligned}\mathscr{H}\psi_2 = {} & H_{22}\cos\theta\varphi_2 + H_{23}\cos\theta\varphi_3 + H_{32}\sin\theta\varphi_2 \\ & + H_{33}\sin\theta\varphi_3 = E_2\varphi_2\cos\theta + E_2\varphi_3\sin\theta\,.\end{aligned} \tag{11}$$

Since φ_2 and φ_3 are mutually orthogonal, this equation implies that the following two equations must be simultaneously fulfilled

$$H_{22}\cos\theta + H_{23}\sin\theta = E_2\cos\theta\,, \tag{12a}$$

$$H_{33}\sin\theta + H_{23}\cos\theta = E_2\sin\theta\,, \tag{12b}$$

where we have used the identity $H_{23} = H_{32}$.

In Eqs. (12a) and (12b) one may eliminate E_2 by multiplying the former by $\sin\theta$ and the latter by $\cos\theta$ and subtracting to obtain

$$(H_{22} - H_{33})\sin 2\theta/2 - H_{23}\cos 2\theta = 0$$

or

$$\tan 2\theta = 2H_{23}/(H_{22} - H_{33})\,. \tag{13}$$

By substituting the values of the matrix elements, one obtains

$$\tan 2\theta = J_{AB}/(\nu_A - \nu_B)\,. \tag{14}$$

In the limit of very weak coupling the off-diagonal element J_{AB} is negligible and $\theta = 0$, so that φ_2 and φ_3 become eigenfunctions belonging to the eigenvalues H_{22} and H_{33} respectively. More generally, θ takes on an intermediate value and E_2 is obtained from function (11) as

$$E_2 = H_{22} + H_{23}\tan\theta\,. \tag{15}$$

In a similar manner one can derive for E_3 the expression

$$E_3 = H_{33} - H_{23}\tan\theta\,. \tag{16}$$

For small values of θ, $\tan\theta \simeq 1/2\tan 2\theta$ and the eigenvalues E_2 and E_3 deviate from the diagonal elements H_{22} and H_{33} by expressions quadratic in J_{AB}.

The relative intensities are

$$I_{12} = |\langle\alpha\alpha|F_x|\alpha\beta\cos\theta + \beta\alpha\sin\theta\rangle|^2 = (1/2)^2(\cos\theta + \sin\theta)^2 = (1+\sin 2\theta)/4\,, \tag{17}$$

$$I_{13} = |\langle\alpha\alpha|F_x|-\alpha\beta\sin\theta + \beta\alpha\cos\theta\rangle|^2 = (1/2)^2(-\sin\theta + \cos\theta)^2 = (1-\sin 2\theta)/4\,, \tag{18}$$

$$I_{24} = I_{12} \quad \text{and} \quad I_{34} = I_{13}\,.$$

The transition frequencies are

$$\nu_{12} = H_{11} - H_{22} - H_{23}\tan\theta\,, \tag{19}$$

$$\nu_{13} = H_{11} - H_{33} + H_{23}\tan\theta\,, \tag{20}$$

$$\nu_{24} = H_{22} - H_{44} + H_{23}\tan\theta\,, \tag{21}$$

$$\nu_{34} = H_{33} - H_{44} - H_{23}\tan\theta\,. \tag{22}$$

By the use of some algebra, these formulae may be seen to be equivalent to those of Section IV.2.b. To see this, we will first write down the expression for the frequency intervals a and b. One obtains

$$a = \nu_{13} - \nu_{34} = \nu_A - \nu_B + J_{AB}(1+\tan\theta)\,, \tag{23a}$$

$$b = \nu_{24} - \nu_{12} = \nu_A - \nu_B - J_{AB}(1-\tan\theta)\,, \tag{23b}$$

and it is immediately seen that

$$\frac{a-b}{2} = J_{AB}\,. \tag{24}$$

For the chemical shift $\Delta = \nu_A - \nu_B$ one obtains the expression

$$ab = [\Delta^2 + 2\Delta J_{AB}\tan\theta - J_{AB}^2(1-\tan^2\theta)]\,.$$

By the use of the equation

$$\tan 2\theta = \frac{2\tan\theta}{1-\tan^2\theta}\,, \tag{25}$$

we may replace the expression $1-\tan^2\theta$ in the last bracket to obtain

$$ab = \Delta^2 + 2\Delta J_{AB}\tan\theta - \frac{J_{AB}^2\, 2\tan\theta}{\tan 2\theta}\,. \tag{26}$$

Finally, $\tan 2\theta$ may be replaced by J_{AB}/Δ to yield

$$ab = \Delta^2\,. \tag{27}$$

In the limit of very strong coupling when $J_{AB}/(\nu_A - \nu_B)$ tends to infinity, the angle θ tends to $\pi/4 = 45°$ and the eigenfunctions become either totally symmetric or totally antisymmetric under exchange of the two spins. The symmetric function corresponds to a nuclear triplet state and yields the only observable transition, while the nuclear singlet state $(\alpha\beta-\beta\alpha)/\sqrt{2}$ which is antisymmetric in the two spins does not couple with the external radiofrequency field and hence does not contribute to the observed spectrum.

d) The Choice of Basis Functions; the Uses of Symmetry

If the Hamiltonian of a spin system is invariant under the exchange of spins, the basis functions of the corresponding spins should be chosen to be eigenfunctions of the corresponding exchange operation. For example, the basis function of an AB_2 spin system should be chosen such that the B spin functions are either symmetric or antisymmetric in the two spins. In more complex spin systems, for example, AA′XX′, the Hamiltonian is not invariant under the exchange of one pair of spins, but the mirror-image symmetry of the molecule does not change the Hamiltonian by a simultaneous permutation of A for A′ and X for X′. A suitable set of basis functions is then a product of symmetrized basis functions for the AA′ and XX′ species. This will be elaborated in the following.

e) An Example: The AB_2 Spin System

The basis set of the three-spin system AB_2 consists of $2^3=8$ basis functions. These may be chosen as the products of an α or β spin function of nucleus A with the symmetrized basis spin functions of the B_2 spins. The basis functions thus may be written

$$\begin{aligned}
&\varphi_1=\alpha\alpha\alpha, && \varphi_5=\beta(\alpha\beta+\beta\alpha)/\sqrt{2},\\
&\varphi_2=\alpha(\alpha\beta+\beta\alpha)/\sqrt{2}, && \varphi_6=\beta\beta\beta,\\
&\varphi_3=\beta\alpha\alpha, && \varphi_7=\alpha(\alpha\beta-\beta\alpha)/\sqrt{2},\\
&\varphi_4=\alpha\beta\beta, && \varphi_8=\beta(\alpha\beta-\beta\alpha)/\sqrt{2}.
\end{aligned}$$

Of the basis functions just enumerated, functions φ_1 through φ_6 are symmetric under an exchange of the two B nuclei, while φ_7 and φ_8 are antisymmetric. As always with simple product functions or symmetrized combinations of such functions, the basis functions are also basis functions of $F_z=\sum_i I_z(i)$. Off-diagonal elements of the Hamiltonian can only occur between basis functions belonging to the same eigenvalue M of F_z and belonging to the same symmetry species. This may be shown to be rigorously true but also follows from the explicit calculation of the matrix elements according to the prescriptions outlined in the AB case above.

The Hamiltonian of the AB_2 spin system is

$$\mathscr{H}=\nu_A I_z(A)+\nu_B F_z(B)+J_{AB}\boldsymbol{I}(A)\cdot\boldsymbol{F}(B) \qquad (28)$$

where $\boldsymbol{F}(B)$ is the resultant spin angular momentum for the two B nuclei. In this Hamiltonian we have omitted a possible coupling between the two B nuclei because this coupling always remains unobservable. The non-vanishing matrix elements of the Hamiltonian are

$$H_{11} = \nu_A/2 + \nu_B + J_{AB}/2,$$
$$H_{22} = \nu_A/2,$$
$$H_{33} = -\nu_A/2 + \nu_B - J_{AB}/2,$$
$$H_{44} = \nu_A/2 - \nu_B - J_{AB}/2,$$
$$H_{55} = -\nu_A/2,$$
$$H_{66} = -\nu_A/2 - \nu_B + J_{AB}/2,$$
$$H_{77} = \nu_A/2,$$
$$H_{88} = -\nu_A/2,$$
$$H_{23} = H_{32} = H_{45} = H_{54} = J_{AB}/\sqrt{2}.$$

From the foregoing list it is evident that φ_1, φ_6, φ_7, and φ_8 are eigenfunctions of the Hamiltonian. However, the functions φ_2 and φ_3 become mixed, as do φ_4 and φ_5, and the corresponding eigenfunctions may be written

$$\psi_2 = \varphi_2 \cos\theta_+ + \varphi_3 \sin\theta_+, \tag{29a}$$
$$\psi_3 = -\varphi_2 \sin\theta_+ + \varphi_3 \cos\theta_+, \tag{29b}$$
$$\psi_4 = \varphi_4 \cos\theta_- + \varphi_5 \sin\theta_-, \tag{29c}$$
$$\psi_5 = -\varphi_4 \sin\theta_- + \varphi_5 \cos\theta_+. \tag{29d}$$

The angles θ_+ and θ_- may be defined by $\tan 2\theta_+ = 2H_{23}/(H_{22}-H_{33})$ and $\tan 2\theta_- = 2H_{45}/(H_{44}-H_{55})$. The corresponding eigenvalues may be written

$$E_2 = H_{22} + H_{23} \tan\theta_+, \tag{30a}$$
$$E_3 = H_{33} - H_{23} \tan\theta_+, \tag{30b}$$
$$E_4 = H_{44} + H_{45} \tan\theta_-, \tag{30c}$$
$$E_5 = H_{55} - H_{45} \tan\theta_-. \tag{30d}$$

The transition frequencies are given by the differences in the corresponding eigenvalues and the relative intensities are given by the following list:

$$I_{12} = (\sqrt{2} \sin\theta_+ - \cos\theta_+)^2, \tag{31a}$$
$$I_{13} = (\sqrt{2} \cos\theta_+ + \sin\theta_+)^2, \tag{31b}$$
$$I_{24} = [\sqrt{2} \cos(\theta_+ - \theta_-) + \cos\theta_+ \sin\theta_-]^2, \tag{31c}$$
$$I_{25} = [\sqrt{2} \sin(\theta_+ - \theta_-) + \cos\theta_+ \cos\theta_-]^2, \tag{31d}$$
$$I_{34} = [\sqrt{2} \sin(\theta_+ - \theta_-) + \sin\theta_+ \sin\theta_-]^2, \tag{31e}$$
$$I_{35} = [\sqrt{2} \cos(\theta_+ - \theta_-) - \sin\theta_+ \cos\theta_-]^2, \tag{31f}$$
$$I_{46} = (\sqrt{2} \sin\theta_- + \cos\theta_-)^2, \tag{31g}$$
$$I_{56} = (\sqrt{2} \cos\theta_- - \sin\theta_-)^2, \tag{31h}$$
$$I_{78} = 1. \tag{31i}$$

It will be noted that the frequency ν_{78} is always centered on the Larmor frequency ν_A of nucleus A.

Apart from a scale factor, the appearance of the spectrum depends only on the single parameter. $J_{AB}/(\nu_A - \nu_B)$ and a list of reduced transition frequencies and relative intensities may be given once and for all for an AB_2 spin system.

f) Rules for Evaluating Matrix Elements

The basis functions employed will normally be either the simple product functions or suitably symmetrized linear combinations of the simple product functions. Since the Hamiltonian is linear, it is sufficient to lay down the rules for the simple product functions themselves, because matrix elements between linear combinations of basic products are evaluated by expansion. The diagonal elements of the Hamiltonian may be written

$$H_{jj} = \langle \varphi_j | \mathscr{H} | \varphi_j \rangle = \sum_k \nu_k m_k + \sum_{k<l} J_{kl} m_k m_l \tag{32}$$

where m_k is the magnetic quantum number of spin k and equals $+1/2$ if spin k has state α, but equals $-1/2$ if spin k has state β. The off-diagonal elements may be written

$$H_{ij} = \langle \varphi_i | \mathscr{H} | \varphi_j \rangle = J_{kl} U_{ij}/2 \tag{33}$$

where U_{ij} equals 1 if φ_i differs from φ_j by an interchange of spin k and l, and is zero otherwise.

If the Hamiltonian is invariant under certain symmetry operations and if the basis functions are chosen to be eigenfunctions of the same symmetry operations, then it is not necessary to evaluate matrix elements between different symmetry species because they must of necessity vanish.

The eigenvalues are obtained by solving the secular equation and the eigenfunctions follow by solution of the set of linear equations. The complete solution of the eigenvalue problem is equivalent to a reduction of the Hamiltonian matrix to diagonal form. The number of possible transitions is limited by the selection rule $\Delta M = \pm 1$, where M is the magnetic quantum number of the eigenfunction and equals the eigenvalue of $F_z = \sum_i I_z(i)$. For symmetrical molecules, transitions may only occur between functions of the same symmetry. The relative intensities are proportional to the square of the matrix element of $F_x = \sum_j I_x(j)$ between the eigenfunctions. For simple product functions, the matrix element of F_x equals 1/2 between products which are identical except for the state of one single spin, and is zero otherwise. Matrix elements for linear combinations of basis product functions are evaluated by expansion.

g) The ABC and ABX Spin Systems

The Hamiltonian of a three-spin system ABC may be written

$$\begin{aligned} \mathscr{H} = {} & \nu_A I_z(A) + \nu_B I_z(B) + \nu_C I_z(C) + J_{AB} \boldsymbol{I}(A) \cdot \boldsymbol{I}(B) \\ & + J_{AC} \cdot \boldsymbol{I}(A) \cdot \boldsymbol{I}(C) + J_{BC} \boldsymbol{I}(B) \cdot \boldsymbol{I}(C) . \end{aligned} \tag{34}$$

Table 7. *Basic functions and diagonal matrix elements in an ABC spin system*

Basis functions	M	Diagonal matrix element, H_{ii}
$\varphi_1=\alpha\alpha\alpha$	3/2	$(\nu_A+\nu_B+\nu_C)/2+(J_{AB}+J_{AC}+J_{BC})/4$
$\varphi_2=\alpha\alpha\beta$	1/2	$(\nu_A+\nu_B-\nu_C)/2+(J_{AB}-J_{AC}-J_{BC})/4$
$\varphi_3=\alpha\beta\alpha$	1/2	$(\nu_A-\nu_B+\nu_C)/2+(-J_{AB}+J_{AC}-J_{BC})/4$
$\varphi_4=\beta\alpha\alpha$	1/2	$(-\nu_A+\nu_B+\nu_C)/2+(-J_{AB}-J_{AC}+J_{BC})/4$
$\varphi_5=\alpha\beta\beta$	−1/2	$(\nu_A-\nu_B-\nu_C)/2+(-J_{AB}-J_{AC}+J_{BC})/4$
$\varphi_6=\beta\alpha\beta$	−1/2	$(-\nu_A+\nu_B-\nu_C)/2+(-J_{AB}+J_{AC}-J_{BC})/4$
$\varphi_7=\beta\beta\alpha$	−1/2	$(-\nu_A-\nu_B+\nu_C)/2+(J_{AB}-J_{AC}-J_{BC})/4$
$\varphi_8=\beta\beta\beta$	−3/2	$(-\nu_A-\nu_B-\nu_C)/2+(J_{AB}+J_{AC}+J_{BC})/4$

The basis functions together with the matrix elements of the Hamiltonian are listed in Table 7.

The non-vanishing off-diagonal elements are

$$H_{23}=H_{67}=J_{BC}/2, \tag{35a}$$

$$H_{34}=H_{56}=J_{AB}/2, \tag{35b}$$

$$H_{24}=H_{57}=J_{AC}/2. \tag{35c}$$

If none of the off-diagonal elements can be neglected, the reduction of the Hamiltonian matrix to diagonal form involves the solution of two third order secular equations and the explicit solutions are not conveniently written in analytical form. The problem is then preferably solved by means of the electronic computer.

In the ABX approximation, nucleus C will be called spin X and all off-diagonal elements excluding J_{AB} are neglected. The problem then reduces to the diagonalization of two 2×2 submatrices involving states φ_3 and φ_4 on the one hand and φ_5 and φ_6 on the other hand. These can be solved analytically by introducing the angles θ_+ and θ_- as follows

$$\psi_3=\varphi_3\cos\theta_+ +\varphi_4\sin\theta_+ , \tag{36a}$$

$$\psi_4=-\varphi_3\sin\theta_+ +\varphi_4\cos\theta_+ , \tag{36b}$$

$$\psi_5=\varphi_5\cos\theta_- +\varphi_6\sin\theta_- , \tag{36c}$$

$$\psi_6=-\varphi_5\sin\theta_- +\varphi_6\cos\theta_- . \tag{36d}$$

The corresponding eigenvalues may be written

$$E_3=H_{33}+H_{34}\tan\theta_+ , \tag{37a}$$

$$E_4=H_{44}-H_{34}\tan\theta_+ , \tag{37b}$$

$$E_5=H_{55}+H_{56}\tan\theta_- , \tag{37c}$$

$$E_6=H_{66}-H_{56}\tan\theta_- , \tag{37d}$$

and the angles are determined by the equations

$$\tan 2\theta_+ =J_{AB}/[(\nu_A-\nu_B)+(J_{AX}-J_{BX})/2], \tag{38a}$$

$$\tan 2\theta_- =J_{AB}/[(\nu_A-\nu_B)-(J_{AX}-J_{BX})/2]. \tag{38b}$$

It will be noted that these two expressions are analogous to those of the AB spin system, except that $\nu_A - \nu_B$ is replaced by the effective Larmor frequencies $\nu_A - \nu_B \pm (J_{AX} - J_{BX})/2$.

This analogy is the formal basis for the subspectral analysis procedure. In the ABX approximation, the basis functions φ_1, φ_2, φ_7, and φ_8 may be considered as eigenfunctions of the Hamiltonian. There are altogether 14 allowed transitions in the ABX spectrum and these are listed in Table 8.

Table 8. *Transition frequencies and relative intensities for an ABX spin system*

No.	Transition	Frequency	Relative intensity
1	8→6	$H_{66} - H_{56} \tan\theta_- - H_{88}$	$1 - \sin 2\theta_-$
2	7→4	$H_{44} - H_{34} \tan\theta_+ - H_{77}$	$1 - \sin 2\theta_+$
3	5→2	$H_{22} - H_{55} - H_{56} \tan\theta_-$	$1 + \sin 2\theta_-$
4	3→1	$H_{11} - H_{33} - H_{34} \tan\theta_+$	$1 + \sin 2\theta_+$
5	8→5	$H_{55} + H_{56} \tan\theta_- - H_{88}$	$1 + \sin 2\theta_-$
6	7→3	$H_{33} + H_{34} \tan\theta_+ - H_{77}$	$1 + \sin 2\theta_+$
7	6→2	$H_{22} - H_{66} + H_{56} \tan\theta_-$	$1 - \sin 2\theta_-$
8	4→1	$H_{11} - H_{44} + H_{34} \tan\theta_+$	$1 - \sin 2\theta_+$
9	8→7	$H_{77} - H_{88}$	1
10	5→3	$H_{33} + H_{34} \tan\theta_+ - H_{55} - H_{56} \tan\theta_-$	$\cos^2(\theta_+ - \theta_-)$
11	6→4	$H_{44} - H_{34} \tan\theta_+ - H_{66} + H_{56} \tan\theta_-$	$\cos^2(\theta_+ - \theta_-)$
12	2→1	$H_{11} - H_{22}$	1
13	5→4	$H_{44} - H_{34} \tan\theta_+ - H_{55} - H_{56} \tan\theta_-$	$\sin^2(\theta_+ - \theta_-)$
14	6→3	$H_{33} + H_{34} \tan\theta_+ - H_{66} + H_{66} \tan\theta_-$	$\sin^2(\theta_+ - \theta_-)$

The values of the diagonal element in Table 8 are those listed in Table 8 and the off-diagonal elements in Table 8 are all equal to $J_{AB}/2$ and the angles θ_+ and θ_- finally are defined in Eqs. (38a) and (38b). The practical analysis procedure for ABX spectra has been discussed in Section IV.4.

h) Two Examples: The AA'XX' and AA'BB' Spin Systems

The Hamiltonian of the symmetrical-form spin systems depends on two chemical shift parameters, ν_A and ν_B, and on four spin-spin coupling parameters.

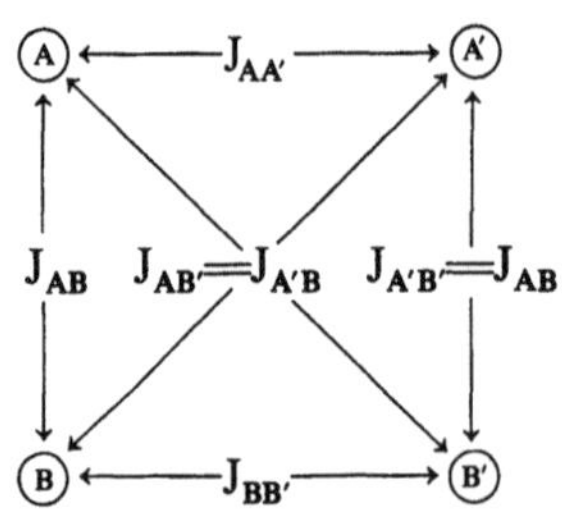

It may be written

$$\mathscr{H} = \nu_A F_z(A) + \nu_B F_z(B) + J_{AA'} \boldsymbol{I}(A) \cdot \boldsymbol{I}(A') + J_{BB'} \boldsymbol{I}(B) \cdot \boldsymbol{I}(B') + J_{AB}[\boldsymbol{I}(A) \cdot \boldsymbol{I}(B) + \boldsymbol{I}(A') \cdot \boldsymbol{I}(B')] + J_{A'B}[\boldsymbol{I}(A') \cdot \boldsymbol{I}(B) + \boldsymbol{I}(A) \cdot \boldsymbol{I}(B')] \,. \tag{39}$$

Table 9. *Basis functions and matrix elements for an AA'BB' spin system*

Basis function	M	Symmetry	Diagonal matrix elements[a]	Non-vanishing off-diagonal matrix elements
$\varphi_1 = \alpha\alpha\alpha\alpha$	2	S	$H_{11} = \nu_A + \nu_B + N/2$	$H_{23} = N/2$
$\varphi_2 = (\alpha\beta + \beta\alpha)\,\alpha\alpha/\sqrt{2}$	1	S	$H_{22} = \nu_B$	$H_{46} = L/2;\ H_{47} = N/2$
$\varphi_3 = \alpha\alpha(\alpha\beta + \beta\alpha)\sqrt{2}$	1	S	$H_{33} = \nu_A$	$H_{56} = L/2;\ H_{57} = N/2$
$\varphi_4 = \beta\beta\alpha\alpha$	0	S	$H_{44} = -\nu_A + \nu_B - N/2$	$H_{67} = -L/2$
$\varphi_5 = \alpha\alpha\beta\beta$	0	S	$H_{55} = \nu_A - \nu_B - N/2$	$H_{8,9} = N/2$
$\varphi_6 = (\alpha\beta - \beta\alpha)(\alpha\beta - \beta\alpha)/2$	0	S	$H_{66} = -K$	
$\varphi_7 = (\alpha\beta + \beta\alpha)(\alpha\beta + \beta\alpha)/2$	0	S	$H_{77} = 0$	
$\varphi_8 = (\alpha\beta + \beta\alpha)\,\beta\beta/\sqrt{2}$	−1	S	$H_{88} = -\nu_B$	
$\varphi_9 = \beta\beta(\alpha\beta + \beta\alpha)/\sqrt{2}$	−1	S	$H_{99} = -\nu_A$	
$\varphi_{10} = \beta\beta\beta\beta$	−2	S	$H_{10,10} = \nu_A - \nu_B + N/2$	$H_{11,12} = -L/2$
$\varphi_{11} = (\alpha\beta - \beta\alpha)\,\alpha\alpha/\sqrt{2}$	1	A	$H_{11,11} = \nu_B - K/2 - M/2$	$H_{13,14} = -L/2$
$\varphi_{12} = \alpha\alpha(\alpha\beta - \beta\alpha)/\sqrt{2}$	1	A	$H_{12,12} = \nu_A - K/2 + M/2$	$H_{15,16} = -L/2$
$\varphi_{13} = (\alpha\beta + \beta\alpha)(\alpha\beta - \beta\alpha)/2$	0	A	$H_{13,13} = -K/2 + M/2$	
$\varphi_{14} = (\alpha\beta - \beta\alpha)(\alpha\beta + \beta\alpha)/2$	0	A	$H_{14,14} = -K/2 - M/2$	
$\varphi_{15} = (\alpha\beta - \beta\alpha)\,\beta\beta/\sqrt{2}$	−1	A	$H_{15,15} = -\nu_B - K/2 - M/2$	
$\varphi_{16} = \beta\beta(\alpha\beta - \beta\alpha)/\sqrt{2}$	−1	A	$H_{16,16} = -\nu_A - K/2 + M/2$	

[a] For convenience, K/2 is substracted from all diagonal elements. This redefines the zero of the energy scale but makes no difference in transition frequencies.

The Hamiltonian remains invariant under the reflection operation in which A is exchanged for A′ concurrently with the exchange B for B′. This mirror-image symmetry makes it advantageous to introduce symmetrized basis functions, and the matrix elements are conveniently expressed in terms of the following linear combinations of the spin-coupling parameters

$$K = J_{AA'} + J_{BB'}, \qquad L = J_{AB} - J_{A'B},$$

$$M = J_{AA'} - J_{BB'}, \qquad N = J_{AB} + J_{A'B}.$$

The basis functions are listed together with the Hamiltonian matrix elements in Table 9.

In a genuine AA′BB′ spin system, all the states φ_4 through φ_7 that belong to the symmetric symmetry species and the magnetic quantum number $M=0$ become mixed through the occurrence of non-vanishing off-diagonal matrix elements. The complete analysis of an AA′BB′ spin system therefore necessitates the diagonalization of a 4×4 submatrix and the solution of a fourth order secular equation. The solution cannot therefore be conveniently given in closed form. In the AA′XX′ limit one may neglect the mixing of states 4 and 5 with states 6 and 7, and the 4×4 submatrix reduces to two diagonal elements and one 2×2 submatrix. Among the antisymmetric states, one may in the AA′XX′ limit neglect the mixing of states 11 and 12 and states 15 and 16, where the diagonal elements differ by the chemical shift difference. The whole diagonalization therefore involves the diagonalization of two 2×2 submatrices. These are listed in Table 10.

Table 10. *Eigen functions and energies in AA′XX′ spin systems*

State	Wave function	Energy
ψ_6	$\varphi_6 \cos\theta_s + \varphi_7 \sin\theta_s$	$-K/2-(K^2+L^2)^{1/2}/2$
ψ_7	$-\varphi_6 \sin\theta_s + \varphi_7 \cos\theta_s$	$-K/2-(K^2+L^2)^{1/2}/2$
ψ_{13}	$\varphi_{13} \cos\theta_a - \varphi_{14} \sin\theta_a$	$-K/2+(M^2+L^2)^{1/2}/2$
ψ_{14}	$\varphi_{13} \sin\theta_a + \varphi_{14} \cos\theta_a$	$-K/2-(M^2+L^2)^{1/2}/2$

Because of the convenient algebraic form, the energy eigenvalues have not been written in terms of the transformation angles θ_a and θ_s which are defined by the relationships

$$\cos 2\theta_s : \sin 2\theta_s : 1 = K : L : (K^2 + L^2)^{1/2}, \tag{40a}$$

$$\cos 2\theta_a : \sin 2\theta_a : 1 = M : L : (M^2 + L^2)^{1/2}. \tag{40b}$$

The spectrum consists of two identical half-spectra, one assigned to the AA′ transitions and the other to the XX′ transitions. Each half-spectrum consist of twelve transitions symmetrically displaced about the corresponding Larmor frequency. They are listed in Table 11.

The practical analysis procedure for an AA′XX′ spectrum in terms of sub-spectra has been described in Section IV.9. In genuine AA′BB′ spectra, it is not possible to write down explicit expressions for the transitions involving states φ_4 through φ_7, and the reader is referred either to the analysis procedure described in Section IV.9. or to a complete numerical diagonalization.

Table 11. *Transition frequencies and relative intensities for a half-spectrum of an AA'XX' spin system. The transition frequencies are listed relative to the Larmor frequencies of the corresponding nucleus*

Transition		Frequency	Relative intensity
1	2→1	N/2	1
2	4→2	N/2	1
3	10→8	−N/2	1
4	8→5	−N/2	1
5	6→3	$K/2+(K^2+L^2)^{1/2}/2$	$\sin^2\theta_s$
6	9→7	$-K/2+(K^2+L^2)^{1/2}/2$	$\cos^2\theta_s$
7	7→3	$+K/2-(K^2+L^2)^{1/2}/2$	$\cos^2\theta_s$
8	9→6	$-K/2-(K^2+L^2)^{1/2}/2$	$\sin^2\theta_s$
9	14→12	$M/2+(M^2+L^2)^{1/2}/2$	$\sin^2\theta_a$
10	16→13	$-M/2+(M^2+L^2)^{1/2}/2$	$\cos^2\theta_a$
11	13→12	$+M/2-(M^2+L^2)^{1/2}/2$	$\cos^2\theta_a$
12	16→14	$-M/2-(M^2+L^2)^{1/2}/2$	$\sin^2\theta_a$

V.3. Iterative and Least Squares Fit Procedures

The emphasis in the present book has been on analysis procedures that do not require access to a digital computer but may be performed by means of a slide rule and/or a desk-top calculator. With complex spin systems, however, one will normally have recourse to numerical procedures and employ the services of a fast digital computer. In the most pedestrian approach the computer is employed solely to diagonalize the Hamiltonian matrix, but there are a number of more sophisticated programs available today which iterate the solution for a least-squares fit between calculated and observed spectra. Most of these programs make use of the fact that a small change in the parameters affects the wave functions in higher order than it affects the eigenvalues. The procedure employed than is as follows. First, a set of trial parameters is devised, preferably as close as possible to the true parameters. The Hamiltonian including these parameters is then diagonalized and the eigenvalues (or transition frequencies) obtained are then compared with the observed ones. The differences obtained are than transformed by means of the calculated eigenfunctions to changes in the NMR parameter to yield a new set of trial parameters. This procedure is repeated – or iterated – until a least-squares fit between calculated and observed transition energies is obtained, or alternatively until the root-mean-square deviation falls below a preassigned value. Unfortunately, no use is made of the intensity information during the iterative procedure. It is customary, however, to complete the comparison between observed and theoretical spectra by calculating not only the transition frequencies but also the relative intensities and comparing the full spectrum with the observed one. The earliest complete computer programs for iterative analysis of NMR spectra were prepared by REILLY and SWALEN (NMREN and NMRIT) and by BOTHNER-BY and CASTELLANO (LAOCOON I and II) [*13*]. These programs have been subjected to several improvements and changes, in part by other scientists. A list of computer programs available for iterative analysis of NMR spectra may be obtained through: Quantum Chemistry Program Exchange, Bloomington, Indiana, USA.

Two warnings should be issued to users of computer programs. First, it may be that the analysis of an NMR spectrum may not yield a unique solution because the spectrum lacks sufficient information. This possibility is not always recognized by the computer, which may provide an excellent least-squares fit and stop at a particular solution which may not necessarily be the correct one. Secondly, it must be emphasized that, if ambiguities exist concerning the relative signs of the spin-coupling constants, then the computer must be free to iterate with any combination of signs. It is *not* sufficient after a best fit has been obtained to change the relative signs and compare a calculated spectrum with the observed one without further iteration.

VI. Historical Notes and Suggestions for Further Reading

The form of the high resolution nuclear spin Hamiltonian was found by HAHN and MAXWELL [14] and by GUTOWSKY, MCCALL, and SLICHTER [15]. HAHN and MAXWELL analysed the AB spin system, and GUTOWSKY, MCCALL, and SLICHTER laid down the rules for simple first order spectra. BANERJEE, DAS, and SAHA [16] wrote a paper on the analysis of the general A_nB_m spin system but unfortunately used a formalism which prevented the paper from being very widely employed. A severe test of the accuracy of the spin Hamiltonian was the comparison between calculated and observed proton NMR spectra on some simple hydrocarbons by ANDERSON [17] and ARNOLD [18]. The composite particle concept is implied in the perturbation treatment developed by ANDERSON [17]. The uses of symmetrized basis functions for spectra containing symmetrically equivalent nuclei and simple rules for evaluating the matrix elements were published by MCCONNELL, MCLEAN, and REILLY [19]. In 1957 POPLE, BERNSTEIN, and SCHNEIDER published a series of epoch-making papers in which they laid down the nomenclature now generally employed for spin systems, summarized the earlier development and contributed, among other things, the analysis procedure for AA'BB' spin systems (cf. ref. 20).

Unfortunately the distinction between symmetric equivalence and magnetic equivalence was not emphasized particularly by these authors, and this subject has been brought up again several times, for example, by WAUGH and DOBBS [21] and by WHITMAN, ONSAGER, SAUNDERS, and DUBB [22]. The general procedure for analyzing spectra with one pair of strongly coupled spins was developed in a series of papers by POPLE and SCHAEFER [23], DIEHL and POPLE [24] and HOFFMAN and GRONOWITZ [2], and the slide-rule procedure for analyzing AB spectra was developed by HOFFMAN and GRONOWITZ [25]. Iterative procedures for analyzing NMR spectra were developed by HOFFMAN and GRONOWITZ [2], by REILLY and SWALEN [26], by ALEXANDER [27] and by BOTHNER-BY and coworkers [28–30]. The equivalence of the method of HOFFMAN and GRONOWITZ on the one hand, and the method of ALEXANDER and REILLY and SWALEN on the other, was demonstrated in a paper by HOFFMAN [31]. The method of subspectra was developed in great detail by DIEHL, HARRIS, and JONES [32]. Review articles on the status of NMR spectra analysis have been written by CORIO [33] and by HOFFMAN [34]. A review article on the use of computer techniques in NMR has been written by SWALEN [13].

For further reading on the subject of analysis of NMR spectra, Chapter 6 of POPLE, BERNSTEIN, and SCHNEIDER's monograph [20] is still one of the best. Simplified and less comprehensive treatments may be found in the book "Nuclear Magnetic Resonance Spectroscopy" by LYNDEN-BELL and HARRIS (Nelson and Sons Ltd., London 1969) and in D. W. MATHIESON (Ed.) "Nuclear Magnetic Resonance for Organic Chemists" Academic Press, London/New York 1966. Most textbooks on NMR contain sections in which the analysis of

NMR spectra is described, for example, EMSLEY, FEENEY, SUTCLIFFE, "High Resolution Nuclear Magnetic Resonance Spectroscopy", Vol. 1 (Pergamon Press, Oxford 1965), and JACKMAN and STERNHELL "Applications of Nuclear Magnetic Resonance Spectroscopy in Organic Chemistry" (Pergamon Press, Oxford 1969), and F.A. BOVEY "Nuclear Magnetic Resonance Spectroscopy" (Academic Press, New York 1969).

Of books devoted entirely to the subject of analysis of NMR spectra, one may mention the elementary introduction by J.D. ROBERTS "An Introduction to the Analysis of Spin-Spin Splitting in High-Resolution Nuclear Magnetic Resonance" (Benjamin, New York 1961) and the very sophisticated treatment by P.L. CORIO "Structure of High-Resolution NMR Spectra" (Academic Press, New York 1967).

Appendix A: Sign Ambiguities in ABPX Spectra

In Section IV.5.d it was stated that the sign ambiguities in ABPX spectral parameters are lifted by arbitrarily assigned a positive sign to (i) $\nu_A - \nu_B$, (ii) $|J_{AX} + J_{BX}| - |J_{AP} + J_{BP}|$, (iii) J_{AB} and (iv) $J_{AX} + J_{BX}$. With these sign conventions, the number of possible assignments of labels 1 through 4 in the experimentally observed *ab* subspectra are reduced to two different sets

$$\bar{\nu}_1 > \bar{\nu}_2 > \bar{\nu}_3 > \bar{\nu}_4 \tag{1}$$

or

$$\bar{\nu}_2 > \bar{\nu}_1 > \bar{\nu}_4 > \bar{\nu}_3 . \tag{2}$$

From Eq. (58) of Section IV.5 it follows that

$$J_{AX} + J_{BX} - (J_{AP} + J_{BP}) = 2(\bar{\nu}_2 - \bar{\nu}_3) \tag{3}$$

so that, in assigning labels 1 to 4 to the four observed *ab* subspectra, one must always see to it that

$$\bar{\nu}_2 > \bar{\nu}_3 . \tag{4}$$

Similarly one obtains from Eq. (58) of Section IV.5 that

$$J_{AX} + J_{BX} + (J_{AP} + J_{BP}) = 2(\bar{\nu}_1 - \bar{\nu}_4) \tag{5}$$

so that one must have

$$\bar{\nu}_1 > \bar{\nu}_4 \tag{6}$$

in order to fulfill condition (iii) above. The sum $J_{AX} + J_{BX}$ is obtained from Eqs. (58e) and (58f) as

$$J_{AX} + J_{BX} = (\bar{\nu}_1 + \bar{\nu}_2 - \bar{\nu}_3 - \bar{\nu}_4) . \tag{7}$$

The condition that $J_{AX} + J_{BX}$ be positive may be written

$$\bar{\nu}_1 + \bar{\nu}_2 - \bar{\nu}_3 - \bar{\nu}_4 > 0 . \tag{8}$$

In combination with the sum rule of Eq. (56), the inequality in Eq. (8) above may be rewritten

$$\bar{\nu}_1 > \bar{\nu}_3 \tag{9}$$

and

$$\bar{\nu}_2 > \bar{\nu}_4 . \tag{10}$$

When the inequalities of Eqs. (4), (6), (9), and (10) are again combined with the sum rule of Eq. (56) in Section IV.5, the only possibilities that remain are those of Eqs. (1) and (2).

Appendix B: Line Frequencies and Relative Intensities in A_nB_m Spectra

The following tables give the transition frequencies and relative intensities for a number of values of $J_{AB}/(\nu_A - \nu_B)$ for the following A_nB_m type spectra: A_2B, A_3B, A_4B, A_2B_2, and A_3B_2. All resonance frequencies are given in units of the relative chemical shift $\nu_A - \nu_B$, where $\nu_A > \nu_B$. The origin is given by ν_B, except for the A_2B_2 case where one half of the symmetrical spectrum is given with $(\nu_A + \nu_B)/2$ as origin. The relative intensities are given in brackets next to or below the corresponding frequency.

The M transitions are the so-called mixed transitions which become forbidden in A_nX_m limit. These have been omitted in the A_3B_2 case.

Line frequencies and relative intensities for the A_2B system

$\frac{J_{AB}}{(\nu_A - \nu_B)}$ \ Line	A1	A2	A3	A4	B1	B2	B3	B4	M
0.00	1.000 (2.00)	1.000 (2.00)	1.000 (2.00)	1.000 (2.00)	0.000 (1.00)	0.000 (1.00)	0.000 (1.00)	0.000 (1.00)	2.000 (0.00)
0.05	1.026 (1.90)	1.025 (1.90)	0.975 (2.10)	0.976 (2.10)	0.049 (1.10)	−0.003 (0.99)	−0.051 (0.90)	0.000 (1.00)	2.003 (0.00)
0.10	1.055 (1.79)	1.050 (1.82)	0.950 (2.21)	0.955 (2.18)	0.095 (1.21)	−0.010 (0.97)	−0.105 (0.82)	0.000 (1.00)	2.010 (0.00)
0.15	1.087 (1.67)	1.073 (1.74)	0.927 (2.33)	0.935 (2.26)	0.138 (1.33)	−0.022 (0.94)	−0.160 (0.74)	0.000 (1.00)	2.022 (0.00)
0.20	1.122 (1.55)	1.096 (1.66)	0.904 (2.45)	0.918 (2.34)	0.178 (1.45)	−0.040 (0.89)	−0.218 (0.66)	0.000 (1.00)	2.040 (0.00)
0.25	1.159 (1.43)	1.118 (1.60)	0.882 (2.57)	0.902 (2.40)	0.216 (1.57)	−0.061 (0.84)	−0.277 (0.60)	0.000 (1.00)	2.061 (0.00)
0.30	1.200 (1.32)	1.138 (1.54)	0.862 (2.68)	0.888 (2.46)	0.250 (1.68)	−0.088 (0.78)	−0.338 (0.54)	0.000 (1.00)	2.088 (0.00)
0.35	1.244 (1.20)	1.156 (1.49)	0.844 (2.80)	0.875 (2.51)	0.281 (1.80)	−0.119 (0.71)	−0.400 (0.49)	0.000 (1.00)	2.119 (0.00)
0.40	1.290 (1.09)	1.173 (1.44)	0.827 (2.90)	0.863 (2.56)	0.310 (1.91)	−0.153 (0.65)	−0.463 (0.44)	0.000 (1.00)	2.153 (0.00)
0.45	1.339 (0.99)	1.189 (1.40)	0.811 (3.01)	0.853 (2.60)	0.336 (2.01)	−0.192 (0.59)	−0.528 (0.40)	0.000 (1.00)	2.192 (0.00)
0.50	1.390 (0.89)	1.203 (1.36)	0.797 (3.10)	0.843 (2.63)	0.360 (2.11)	−0.233 (0.53)	−0.593 (0.37)	0.000 (1.00)	2.233 (0.01)
0.55	1.444 (0.81)	1.215 (1.33)	0.785 (3.19)	0.834 (2.66)	0.381 (2.19)	−0.278 (0.48)	−0.659 (0.34)	0.000 (1.00)	2.278 (0.01)
0.60	1.500 (0.73)	1.226 (1.30)	0.774 (3.26)	0.826 (2.69)	0.400 (2.27)	−0.326 (0.43)	−0.726 (0.31)	0.000 (1.00)	2.326 (0.01)
0.65	1.558 (0.66)	1,236 (1.27)	0.764 (3.33)	0.819 (2.72)	0.417 (2.34)	−0.377 (0.38)	−0.794 (0.28)	0.000 (1.00)	2.377 (0.01)
0.70	1.617 (0.59)	1.245 (1.25)	0.755 (3.40)	0.812 (2.74)	0.433 (2.41)	−0.429 (0.34)	−0.862 (0.26)	0.000 (1.00)	2.429 (0.01)
0.75	1.678 (0.54)	1.253 (1.23)	0.747 (3.45)	0.806 (2.76)	0.447 (2.46)	−0.484 (0.31)	−0.931 (0.24)	0.000 (1.00)	2.484 (0.01)
0.80	1.740 (0.48)	1.260 (1.21)	0.740 (3.50)	0.800 (2.78)	0.460 (2.52)	−0.540 (0.28)	−1.000 (0.22)	0.000 (1.00)	2.540 (0.01)
0.85	1.804 (0.44)	1.266 (1.19)	0.734 (3.55)	0.795 (2.79)	0.471 (2.56)	−0.598 (0.25)	−1.070 (0.21)	0.000 (1.00)	2.598 (0.01)
0.90	1.868 (0.40)	1.271 (1.18)	0.729 (3.59)	0.790 (2.81)	0.482 (2.60)	−0.658 (0.22)	−1.140 (0.19)	0.000 (1.00)	2.658 (0.01)
0.95	1.934 (0.36)	1.276 (1.16)	0.724 (3.62)	0.785 (2.82)	0.491 (2.64)	−0.719 (0.20)	−1.210 (0.18)	0.000 (1.00)	2.719 (0.01)
1.00	2.000 (0.33)	1.281 (1.15)	0.719 (3.65)	0.781 (2.83)	0.500 (2.67)	−0.781 (0.18)	−1.281 (0.17)	0.000 (1.00)	2.781 (0.01)
2.00	3.414 (0.09)	1.318 (1.05)	0.682 (3.90)	0.732 (2.94)	0.586 (2.91)	−2.146 (0.04)	−2.732 (0.06)	0.000 (1.00)	4.146 (0.01)
3.00	4.886 (0.04)	1.326 (1.02)	0.674 (3.96)	0.712 (2.97)	0.614 (2.96)	−3.598 (0.02)	−4.212 (0.03)	0.000 (1.00)	5.598 (0.01)
4.00	6.372 (0.02)	1.329 (1.01)	0.671 (3.98)	0.702 (2.98)	0.628 (2.98)	−5.074 (0.01)	−5.702 (0.02)	0.000 (1.00)	7.074 (0.00)
5.00	7.864 (0.01)	1.331 (1.01)	0.669 (3.98)	0.695 (2.99)	0.636 (2.99)	−6.559 (0.01)	−7.195 (0.01)	0.000 (1.00)	8.559 (0.00)
10.00	15.348 (0.00)	1.333 (1.00)	0.667 (4.00)	0.681 (3.00)	0.652 (3.00)	−14.030 (0.00)	−14.681 (0.00)	0.000 (1.00)	16.030 (0.00)
∞	—	1.333 (1.00)	0.667 (4.00)	0.667 (3.00)	0.667 (3.00)	—	—	0.000 (1.00)	—

Line frequencies and relative intensities for the A_3B spin system

Line \ $\frac{J_{AB}}{(\nu_A - \nu_B)}$	0.00	0.05	0.10	0.15	0.20	0.25	0.30	0.35	0.40
A1	1.000	1.027	1.058	1.094	1.136	1.183	1.236	1.294	1.358
	(3.00)	(2.84)	(2.65)	(2.45)	(2.23)	(2.00)	(1.77)	(1.55)	(1.35)
A2	1.000	1.026	1.052	1.078	1.103	1.126	1.147	1.166	1.182
	(4.00)	(3.80)	(3.61)	(3.42)	(3.26)	(3.10)	(2.96)	(2.84)	(2.73)
A3	1.000	1.024	1.047	1.067	1.086	1.102	1.117	1.130	1.141
	(3.00)	(2.86)	(2.74)	(2.64)	(2.56)	(2.49)	(2.43)	(2.38)	(2.33)
A4	1.000	0.974	0.948	0.922	0.897	0.874	0.853	0.834	0.818
	(3.00)	(3.16)	(3.35)	(3.55)	(3.77)	(4.00)	(4.22)	(4.44)	(4.64)
A5	1.000	0.976	0.953	0.933	0.914	0.898	0.883	0.870	0.859
	(4.00)	(4.20)	(4.39)	(4.57)	(4.74)	(4.89)	(5.03)	(5.14)	(5.25)
A6	1.000	0.977	0.957	0.939	0.925	0.911	0.900	0.890	0.881
	(3.00)	(3.14)	(3.26)	(3.36)	(3.44)	(3.51)	(3.57)	(3.62)	(3.66)
A7	1.000	1.026	1.052	1.081	1.110	1.140	1.172	1.205	1.239
	(2.00)	(1.90)	(1.80)	(1.70)	(1.61)	(1.51)	(1.43)	(1.34)	(1.26)
A8	1.000	0.976	0.952	0.931	0.910	0.890	0.872	0.855	0.839
	(2.00)	(2.10)	(2.20)	(2.30)	(2.39)	(2.49)	(2.57)	(2.66)	(2.74)
B1	0.000	0.073	0.142	0.206	0.264	0.317	0.364	0.406	0.442
	(1.00)	(1.16)	(1.35)	(1.55)	(1.77)	(2.00)	(2.23)	(2.45)	(2.65)
B2	0.000	0.021	0.032	0.034	0.026	0.008	−0.019	−0.055	−0.099
	(1.00)	(1.04)	(1.05)	(1.03)	(0.97)	(0.90)	(0.81)	(0.71)	(0.61)
B3	0.000	−0.029	−0.067	−0.112	−0.163	−0.220	−0.283	−0.350	−0.421
	(1.00)	(0.94)	(0.87)	(0.78)	(0.70)	(0.62)	(0.54)	(0.48)	(0.42)
B4	0.000	−0.077	−0.157	−0.239	−0.325	−0.411	−0.500	−0.590	−0.681
	(1.00)	(0.86)	(0.74)	(0.64)	(0.56)	(0.49)	(0.43)	(0.38)	(0.34)
B5	0.000	0.024	0.048	0.069	0.090	0.110	0.128	0.145	0.161
	(2.00)	(2.10)	(2.20)	(2.30)	(2.39)	(2.49)	(2.57)	(2.66)	(2.74)
B6	0.000	−0.026	−0.052	−0.081	−0.110	−0.140	−0.172	−0.205	−0.239
	(2.00)	(1.90)	(1.80)	(1.70)	(1.61)	(1.51)	(1.43)	(1.34)	(1.26)
M1	2.000	1.979	1.968	1.966	1.974	1.992	2.019	2.055	2.099
	(0.00)	(0.00)	(0.00)	(0.00)	(0.00)	(0.00)	(0.01)	(0.01)	(0.02)
M2	2.000	2.029	2.067	2.112	2.163	2.220	2.283	2.350	2.421
	(0.00)	(0.00)	(0.00)	(0.00)	(0.00)	(0.00)	(0.00)	(0.00)	(0.00)

Line frequencies and relative intensities for the A_3B spin system

Line \ $\frac{J_{AB}}{(\nu_A - \nu_B)}$	0.45	0.50	0.55	0.60	0.65	0.70	0.75	0.80	0.85
A1	1.427	1.500	1.577	1.657	1.739	1.825	1.911	2.000	2.090
	(1.16)	(1.00)	(0.86)	(0.74)	(0.64)	(0.56)	(0.49)	(0.43)	(0.38)
A2	1.196	1.207	1.217	1.224	1.231	1.236	1.240	1.243	1.246
	(2.64)	(2.56)	(2.49)	(2.43)	(2.38)	(2.34)	(2.30)	(2.27)	(2.24)
A3	1.150	1.159	1.166	1.173	1.179	1.184	1.188	1.192	1.196
	(2.30)	(2.26)	(2.24)	(2.21)	(2.19)	(2.17)	(2.16)	(2.15)	(2.13)
A4	0.804	0.793	0.783	0.776	0.769	0.764	0.760	0.757	0.754
	(4.82)	(4.97)	(5.11)	(5.23)	(5.32)	(5.41)	(5.47)	(5.53)	(5.58)
A5	0.850	0.841	0.834	0.827	0.821	0.816	0.812	0.808	0.804
	(5.33)	(5.41)	(5.48)	(5.53)	(5.58)	(5.62)	(5.66)	(5.69)	(5.72)
A6	0.873	0.866	0.860	0.854	0.849	0.844	0.840	0.836	0.832
	(3.70)	(3.73)	(3.76)	(3.78)	(3.80)	(3.82)	(3.84)	(3.85)	(3.86)
A7	1.273	1.309	1.346	1.383	1.421	1.460	1.500	1.540	1.581
	(1.18)	(1.11)	(1.04)	(0.97)	(0.91)	(0.85)	(0.80)	(0.75)	(0.70)
A8	0.823	0.809	0.796	0.783	0.771	0.760	0.750	0.740	0.731
	(2.82)	(2.89)	(2.96)	(3.03)	(3.09)	(3.15)	(3.20)	(3.25)	(3.30)
B1	0.473	0.500	0.523	0.543	0.561	0.576	0.589	0.600	0.610
	(2.84)	(3.00)	(3.14)	(3.26)	(3.36)	(3.44)	(3.51)	(3.57)	(3.62)
B2	−0.150	−0.207	−0.270	−0.338	−0.410	−0.485	−0.563	−0.643	−0.726
	(0.52)	(0.44)	(0.37)	(0.31)	(0.26)	(0.22)	(0.19)	(0.16)	(0.14)
B3	−0.496	−0.573	−0.653	−0.735	−0.819	−0.904	−0.991	−1.079	−1.168
	(0.37)	(0.32)	(0.28)	(0.25)	(0.22)	(0.20)	(0.18)	(0.16)	(0.14)
B4	−0.773	−0.866	−0.960	−1.054	−1.149	−1.244	−1.340	−1.436	−1.532
	(0.30)	(0.27)	(0.24)	(0.22)	(0.20)	(0.18)	(0.16)	(0.15)	(0.14)
B5	0.177	0.191	0.204	0.217	0.229	0.240	0.250	0.260	0.269
	(2.82)	(2.89)	(2.96)	(3.03)	(3.09)	(3.15)	(3.20)	(3.25)	(3.30)
B6	−0.273	−0.309	−0.346	−0.383	−0.421	−0.460	−0.500	−0.540	−0.581
	(1.18)	(1.11)	(1.04)	(0.97)	(0.91)	(0.85)	(0.80)	(0.75)	(0.70)
M1	2.150	2.207	2.270	2.338	2.410	2.485	2.563	2.643	2.726
	(0.02)	(0.03)	(0.03)	(0.03)	(0.03)	(0.04)	(0.04)	(0.04)	(0.04)
M2	2.496	2.573	2.653	2.735	2.819	2.904	2.991	3.079	3.168
	(0.00)	(0.00)	(0.00)	(0.00)	(0.01)	(0.01)	(0.01)	(0.01)	(0.01)

Line frequencies and relative intensities for the A_3B spin system

Line \ $\frac{J_{AB}}{(\nu_A-\nu_B)}$	0.90	0.95	1.00	2.00	3.00	4.00	5.00	10.00	∞
A1	2.181	2.273	2.366	4.303	6.284	8.275	10.270	20.260	—
	(0.34)	(0.30)	(0.27)	(0.06)	(0.02)	(0.01)	(0.01)	(0.00)	
A2	1.249	1.250	1.252	1.259	1.258	1.256	1.255	1.253	1.250
	(2.21)	(2.19)	(2.18)	(2.04)	(2.02)	(2.01)	(2.01)	(2.00)	(2.00)
A3	1.199	1.302	1.205	1.230	1.237	1.241	1.243	1.247	1.250
	(2.12)	(2.11)	(2.10)	(2.03)	(2.02)	(2.01)	(2.01)	(2.00)	(2.00)
A4	0.751	0.750	0.748	0.741	0.743	0.744	0.745	0.747	0.750
	(5.63)	(5.66)	(5.70)	(5.93)	(5.97)	(5.98)	(5.99)	(6.00)	(6.00)
A5	0.801	0.799	0.795	0.770	0.763	0.759	0.757	0.753	0.750
	(5.74)	(5.76)	(5.78)	(5.94)	(5.97)	(5.98)	(5.99)	(6.00)	(6.00)
A6	0.829	0.826	0.823	0.791	0.779	0.772	0.768	0.759	0.750
	(3.87)	(3.88)	(3.89)	(3.96)	(3.98)	(3.99)	(3.99)	(4.00)	(4.00)
A7	1.623	1.665	1.707	2.618	3.581	4.562	5.550	10.525	—
	(0.66)	(0.62)	(0.59)	(0.21)	(0.10)	(0.06)	(0.04)	(0.01)	
A8	0.723	0.715	0.707	0.618	0.581	0.562	0.550	0.525	0.500
	(3.34)	(3.38)	(3.41)	(3.79)	(3.90)	(3.94)	(3.96)	(3.99)	(4.00)
B1	0.619	0.627	0.634	0.697	0.716	0.725	0.730	0.740	0.750
	(3.66)	(3.70)	(3.73)	(3.94)	(3.98)	(3.99)	(3.99)	(4.00)	(4.00)
B2	−0.811	−0.897	−0.984	−2.864	−4.825	−6.806	−8.795	−18.772	—
	(0.12)	(0.10)	(0.09)	(0.02)	(0.01)	(0.00)	(0.00)	(0.00)	
B3	−1.258	−1.349	−1.441	−3.353	−5.320	−7.303	−9.293	−19.272	—
	(0.13)	(0.12)	(0.11)	(0.03)	(0.01)	(0.01)	(0.00)	(0.00)	
B4	−1.629	−1.726	−1.823	−3.791	−5.779	−7.772	−9.768	−19.759	—
	(0.13)	(0.12)	(0.11)	(0.04)	(0.02)	(0.01)	(0.01)	(0.00)	
B5	0.277	0.285	0.293	0.382	0.419	0.438	0.450	0.475	0.500
	(3.34)	(3.38)	(3.41)	(3.79)	(3.90)	(3.94)	(3.96)	(3.99)	(4.00)
B6	−0.623	−0.665	−0.707	−1.618	−2.581	−3.562	−4.550	−9.525	—
	(0.66)	(0.62)	(0.59)	(0.21)	(0.10)	(0.06)	(0.04)	(0.01)	
M1	2.811	2.897	2.984	4.864	6.825	8.806	10.795	20.772	—
	(0.04)	(0.04)	(0.04)	(0.02)	(0.01)	(0.01)	(0.00)	(0.00)	
M2	3.258	3.349	3.441	5.353	7.320	9.303	11.293	21.272	—
	(0.01)	(0.01)	(0.01)	(0.00)	(0.00)	(0.00)	(0.00)	(0.00)	

The A line frequencies (relative to $(\nu_A + \nu_B)/2$) and relative intensities for the A_2B_2 spin system. The spectrum is symmetrical in $(\nu_A + \nu_B)/2$

$J_{AB}/(\nu_A - \nu_B)$ \ Line	A1	A2	A3	A4	A5	A6	A7	M1	M2
0.00	0.500 (2.00)	0.500 (2.00)	0.500 (2.00)	0.500 (2.00)	0.500 (2.00)	0.500 (2.00)	0.500 (4.00)	1.500 (0.00)	1.500 (0.00)
0.05	0.552 (1.80)	0.550 (1.81)	0.503 (1.98)	0.502 (1.99)	0.450 (2.21)	0.452 (2.20)	0.500 (4.00)	1.455 (0.00)	1.555 (0.00)
0.10	0.610 (1.61)	0.599 (1.66)	0.512 (1.93)	0.508 (1.95)	0.401 (2.46)	0.410 (2.39)	0.500 (4.00)	1.421 (0.00)	1.619 (0.00)
0.15	0.672 (1.43)	0.647 (1.52)	0.529 (1.83)	0.515 (1.91)	0.354 (2.74)	0.372 (2.57)	0.500 (4.00)	1.398 (0.00)	1.691 (0.00)
0.20	0.739 (1.26)	0.694 (1.40)	0.554 (1.70)	0.523 (1.86)	0.310 (3.04)	0.339 (2.74)	0.500 (4.00)	1.387 (0.00)	1.771 (0.00)
0.25	0.809 (1.11)	0.740 (1.30)	0.588 (1.54)	0.530 (1.81)	0.269 (3.36)	0.309 (2.89)	0.500 (4.00)	1.387 (0.00)	1.858 (0.00)
0.30	0.883 (0.97)	0.785 (1.21)	0.631 (1.35)	0.535 (1.76)	0.233 (3.67)	0.283 (3.03)	0.500 (4.00)	1.399 (0.01)	1.951 (0.00)
0.35	0.960 (0.85)	0.828 (1.13)	0.684 (1.17)	0.537 (1.72)	0.202 (3.98)	0.260 (3.15)	0.500 (4.00)	1.422 (0.01)	2.049 (0.00)
0.40	1.040 (0.75)	0.871 (1.06)	0.744 (1.00)	0.536 (1.68)	0.175 (4.25)	0.240 (3.25)	0.500 (4.00)	1.456 (0.01)	2.152 (0.00)
0.45	1.123 (0.66)	0.913 (0.99)	0.812 (0.84)	0.533 (1.65)	0.152 (4.50)	0.223 (3.34)	0.500 (4.00)	1.498 (0.02)	2.259 (0.00)
0.50	1.207 (0.59)	0.955 (0.93)	0.886 (0.71)	0.528 (1.63)	0.134 (4.71)	0.207 (3.41)	0.500 (4.00)	1.548 (0.02)	2.369 (0.00)
0.55	1.293 (0.52)	0.996 (0.88)	0.966 (0.59)	0.521 (1.62)	0.118 (4.89)	0.193 (3.48)	0.500 (4.00)	1.604 (0.02)	2.482 (0.00)
0.60	1.381 (0.46)	1.036 (0.83)	1.049 (0.50)	0.513 (1.61)	0.105 (5.04)	0.181 (3.54)	0.500 (4.00)	1.667 (0.02)	2.599 (0.00)
0.65	1.470 (0.41)	1.077 (0.79)	1.137 (0.42)	0.503 (1.60)	0.094 (5.16)	0.170 (3.59)	0.500 (4.00)	1.734 (0.03)	2.717 (0.00)
0.70	1.560 (0.37)	1.118 (0.75)	1.227 (0.36)	0.493 (1.60)	0.085 (5.27)	0.160 (3.63)	0.500 (4.00)	1.805 (0.03)	2.838 (0.00)
0.75	1.651 (0.34)	1.158 (0.71)	1.320 (0.31)	0.483 (1.60)	0.077 (5.35)	0.151 (3.66)	0.500 (4.00)	1.880 (0.03)	2.961 (0.00)
0.80	1.743 (0.30)	1.199 (0.67)	1.415 (0.27)	0.472 (1.60)	0.070 (5.43)	0.143 (3.70)	0.500 (4.00)	1.957 (0.03)	3.086 (0.00)
0.85	1.836 (0.28)	1.240 (0.64)	1.511 (0.23)	0.461 (1.61)	0.065 (5.49)	0.136 (3.72)	0.500 (4.00)	2.037 (0.03)	3.212 (0.00)
0.90	1.930 (0.25)	1.281 (0.61)	1.609 (0.20)	0.450 (1.62)	0.060 (5.54)	0.130 (3.75)	0.500 (4.00)	2.119 (0.02)	3.340 (0.00)
0.95	2.024 (0.23)	1.322 (0.58)	1.707 (0.18)	0.440 (1.63)	0.058 (5.59)	0.124 (3.77)	0.500 (4.00)	2.203 (0.02)	3.469 (0.00)
1.00	2.118 (0.21)	1.363 (0.55)	1.807 (0.16)	0.429 (1.64)	0.052 (5.63)	0.118 (3.79)	0.500 (4.00)	2.288 (0.02)	3.599 (0.00)
2.00	4.062 (0.06)	2.236 (0.24)	3.848 (0.03)	0.275 (1.81)	0.022 (5.91)	0.062 (3.94)	0.500 (4.00)	4.145 (0.01)	6.359 (0.00)
3.00	6.041 (0.03)	3.169 (0.13)	5.887 (0.01)	0.196 (1.90)	0.014 (5.96)	0.041 (3.97)	0.500 (4.00)	6.097 (0.00)	9.251 (0.00)
4.00	8.031 (0.02)	4.130 (0.08)	7.912 (0.01)	0.151 (1.94)	0.011 (5.98)	0.031 (3.98)	0.500 (4.00)	8.073 (0.00)	12.192 (0.00)
5.00	10.025 (0.01)	5.106 (0.05)	9.928 (0.01)	0.122 (1.96)	0.008 (5.99)	0.025 (3.99)	0.500 (4.00)	10.058 (0.00)	15.155 (0.00)
10.00	20.012 (0.00)	10.054 (0.01)	19.963 (0.00)	0.062 (1.99)	0.004 (6.00)	0.012 (4.00)	0.500 (4.00)	20.029 (0.00)	30.079 (0.00)
∞	—	—	—	0.000 (2.00)	0.000 (6.00)	0.00 (4.00)	0.500 (4.00)	—	—

Line frequencies and relative intensities for the A_4B spin system

Line \ $\frac{J_{AB}}{(\nu_A - \nu_B)}$	0.00	0.05	0.10	0.15	0.20	0.25	0.30	0.35	0.40
A1	1.000	1.028	1.062	1.103	1.153	1.213	1.282	1.360	1.447
	(4.00)	(3.78)	(3.50)	(3.18)	(2.81)	(2.42)	(2.04)	(1.69)	(1.38)
A2	1.000	1.026	1.054	1.082	1.109	1.134	1.155	1.172	1.185
	(6.00)	(5.69)	(5.37)	(5.06)	(4.76)	(4.49)	(4.26)	(4.05)	(3.88)
A3	1.000	1.025	1.049	1.070	1.090	1.106	1.121	1.132	1.142
	(6.00)	(5.71)	(5.46)	(5.23)	(5.04)	(4.89)	(4.75)	(4.64)	(4.55)
A4	1.000	1.024	1.045	1.063	1.078	1.091	1.102	1.112	1.120
	(4.00)	(3.82)	(3.68)	(3.57)	(3.48)	(3.41)	(3.35)	(3.30)	(3.26)
A5	1.000	0.974	0.946	0.918	0.891	0.866	0.845	0.828	0.815
	(4.00)	(4.22)	(4.50)	(4.82)	(5.19)	(5.57)	(5.94)	(6.28)	(6.57)
A6	1.000	0.975	0.951	0.930	0.910	0.894	0.879	0.868	0.858
	(6.00)	(6.31)	(6.63)	(6.94)	(7.23)	(7.49)	(7.72)	(7.91)	(8.07)
A7	1.000	0.976	0.955	0.937	0.922	0.909	0.898	0.888	0.880
	(6.00)	(6.29)	(6.54)	(6.77)	(6.95)	(7.11)	(7.24)	(7.35)	(7.43)
A8	1.000	0.977	0.959	0.943	0.930	0.919	0.910	0.901	0.894
	(4.00)	(4.18)	(4.32)	(4.43)	(4.52)	(4.59)	(4.65)	(4.70)	(4.74)
A9	1.000	1.026	1.055	1.087	1.122	1.159	1.200	1.244	1.290
	(6.00)	(5.69)	(5.36)	(5.01)	(4.66)	(4.30)	(3.95)	(3.60)	(3.28)
A10	1.000	1.025	1.050	1.073	1.096	1.118	1.138	1.156	1.173
	(6.00)	(5.71)	(5.45)	(5.21)	(4.99)	(4.79)	(4.62)	(4.46)	(4.32)
A11	1.000	0.975	0.950	0.927	0.904	0.882	0.862	0.844	0.827
	(6.00)	(6.31)	(6.64)	(6.99)	(7.34)	(7.70)	(8.05)	(8.39)	(8.71)
A12	1.000	0.976	0.955	0.935	0.918	0.902	0.888	0.875	0.863
	(6.00)	(6.29)	(6.55)	(6.79)	(7.01)	(7.20)	(7.38)	(7.53)	(7.67)
B1	0.000	0.097	0.188	0.272	0.347	0.412	0.468	0.515	0.553
	(1.00)	(1.22)	(1.50)	(1.82)	(2.19)	(2.58)	(2.96)	(3.31)	(3.62)
B2	0.000	0.043	0.073	0.087	0.085	0.066	0.031	−0.018	−0.080
	(1.00)	(1.09)	(1.13)	(1.12)	(1.05)	(0.93)	(0.78)	(0.64)	(0.51)
B3	0.000	−0.007	−0.030	−0.066	−0.114	−0.174	−0.244	−0.322	−0.407
	(1.00)	(0.98)	(0.92)	(0.83)	(0.72)	(0.62)	(0.52)	(0.44)	(0.37)
B4	0.000	−0.056	−0.123	−0.199	−0.282	−0.372	−0.467	−0.566	−0.669
	(1.00)	(0.89)	(0.78)	(0.67)	(0.57)	(0.48)	(0.41)	(0.35)	(0.30)
B5	0.000	−0.102	−0.209	−0.318	−0.430	−0.544	−0.660	−0.776	−0.894
	(1.00)	(0.82)	(0.68)	(0.57)	(0.48)	(0.41)	(0.35)	(0.30)	(0.26)
B6	0.000	0.049	0.095	0.138	0.178	0.216	0.250	0.281	0.310
	(3.00)	(3.31)	(3.64)	(3.99)	(4.34)	(4.70)	(5.05)	(5.40)	(5.72)
B7	0.000	−0.003	−0.010	−0.022	−0.040	−0.061	−0.088	−0.119	−0.153
	(3.00)	(2.98)	(2.91)	(2.81)	(2.67)	(2.51)	(2.33)	(2.14)	(1.95)
B8	0.000	−0.051	−0.105	−0.160	−0.218	−0.277	−0.338	−0.400	−0.463
	(3.00)	(2.71)	(2.45)	(2.21)	(1.99)	(1.80)	(1.62)	(1.47)	(1.33)
B9	0.000	0.000	0.000	0.000	0.000	0.000	0.000	0.000	0.000
	(2.00)	(2.00)	(2.00)	(2.00)	(2.00)	(2.00)	(2.00)	(2.00)	(2.00)
M1	2.000	1.957	1.927	1.913	1.915	1.934	1.969	2.018	2.080
	(0.00)	(0.00)	(0.00)	(0.00)	(0.00)	(0.01)	(0.02)	(0.03)	(0.04)
M2	2.000	2.007	2.030	2.066	2.114	2.174	2.244	2.322	2.407
	(0.00)	(0.00)	(0.00)	(0.00)	(0.00)	(0.00)	(0.01)	(0.01)	(0.01)
M3	2.000	2.056	2.123	2.199	2.282	2.372	2.467	2.566	2.669
	(0.00)	(0.00)	(0.00)	(0.00)	(0.00)	(0.00)	(0.00)	(0.00)	(0.00)
M4	2.000	2.003	2.010	2.022	2.040	2.061	2.088	2.119	2.153
	(0.00)	(0.00)	(0.00)	(0.00)	(0.00)	(0.00)	(0.00)	(0.01)	(0.01)

Line frequencies and relative intensities for the A_4B system

Line \ $\frac{J_{AB}}{(\nu_A - \nu_B)}$	0.45	0.50	0.55	0.60	0.65	0.70	0.75	0.80	0.85	0.90
A1	1.541	1.640	1.744	1.852	1.963	2.075	2.190	2.306	2.424	2.542
	(1.13)	(0.92)	(0.76)	(0.63)	(0.53)	(0.45)	(0.38)	(0.33)	(0.29)	(0.25)
A2	1.195	1.203	1.208	1.212	1.215	1.216	1.218	1.218	1.219	1.219
	(3.73)	(3.61)	(3.52)	(3.44)	(3.37)	(3.32)	(3.28)	(3.24)	(3.22)	(3.19)
A3	1.150	1.157	1.162	1.167	1.171	1.174	1.177	1.179	1.182	1.183
	(4.48)	(4.42)	(4.36)	(4.32)	(4.28)	(4.25)	(4.22)	(4.20)	(4.18)	(4.16)
A4	1.127	1.133	1.138	1.143	1.147	1.150	1.153	1.156	1.158	1.161
	(3.23)	(3.20)	(3.18)	(3.16)	(3.14)	(3.13)	(3.12)	(3.11)	(3.10)	(3.09)
A5	0.805	0.797	0.792	0.788	0.785	0.784	0.782	0.782	0.781	0.781
	(6.82)	(7.02)	(7.18)	(7.30)	(7.41)	(7.49)	(7.56)	(7.61)	(7.66)	(7.70)
A6	0.850	0.843	0.838	0.833	0.829	0.826	0.823	0.821	0.818	0.817
	(8.20)	(8.31)	(8.41)	(8.48)	(8.54)	(8.60)	(8.64)	(8.68)	(8.71)	(8.74)
A7	0.873	0.867	0.862	0.857	0.853	0.850	0.847	0.844	0.842	0.839
	(7.51)	(7.57)	(7.62)	(7.66)	(7.70)	(7.73)	(7.76)	(7.78)	(7.80)	(7.82)
A8	0.888	0.883	0.878	0.874	0.870	0.866	0.863	0.860	0.858	0.855
	(4.77)	(4.79)	(4.82)	(4.84)	(4.85)	(4.87)	(4.88)	(4.89)	(4.90)	(4.91)
A9	1.339	1.390	1.444	1.500	1.558	1.617	1.678	1.740	1.804	1.868
	(2.97)	(2.68)	(2.42)	(2.18)	(1.97)	(1.78)	(1.61)	(1.45)	(1.32)	(1.20)
A10	1.189	1.203	1.215	1.226	1.236	1.245	1.253	1.260	1.266	1.271
	(4.20)	(4.09)	(3.99)	(3.90)	(3.82)	(3.75)	(3.69)	(3.63)	(3.58)	(3.53)
A11	0.811	0.797	0.785	0.774	0.764	0.755	0.747	0.740	0.734	0.729
	(9.02)	(9.30)	(9.56)	(9.79)	(10.00)	(10.19)	(10.36)	(10.51)	(10.64)	(10.76)
A12	0.853	0.843	0.834	0.826	0.819	0.812	0.806	0.800	0.795	0.790
	(7.79)	(7.89)	(7.99)	(8.08)	(8.15)	(8.22)	(8.28)	(8.33)	(8.38)	(8.43)
B1	0.584	0.610	0.631	0.648	0.662	0.675	0.685	0.694	0.701	0.708
	(3.87)	(4.08)	(4.24)	(4.37)	(4.47)	(4.55)	(4.62)	(4.67)	(4.71)	(4.75)
B2	−0.152	−0.233	−0.322	−0.416	−0.515	−0.617	−0.723	−0.831	−0.941	−1.053
	(0.40)	(0.31)	(0.24)	(0.19)	(0.15)	(0.12)	(0.10)	(0.08)	(0.07)	(0.06)
B3	−0.498	−0.593	−0.692	−0.795	−0.900	−1.008	−1.118	−1.229	−1.342	−1.455
	(0.31)	(0.26)	(0.22)	(0.18)	(0.16)	(0.14)	(0.12)	(0.10)	(0.09)	(0.08)
B4	−0.775	−0.883	−0.993	−1.105	−1.218	−1.332	−1.448	−1.564	−1.682	−1.799
	(0.26)	(0.23)	(0.20)	(0.17)	(0.15)	(0.14)	(0.12)	(0.11)	(0.10)	(0.09)
B5	−1.013	−1.133	−1.253	−1.374	−1.495	−1.616	−1.738	−1.860	−1.983	−2.105
	(0.23)	(0.21)	(0.18)	(0.16)	(0.15)	(0.13)	(0.12)	(0.11)	(0.10)	(0.09)
B6	0.336	0.360	0.381	0.400	0.417	0.433	0.447	0.460	0.471	0.482
	(6.03)	(6.32)	(6.58)	(6.82)	(7.03)	(7.22)	(7.39)	(7.55)	(7.68)	(7.80)
B7	−0.192	−0.233	−0.278	−0.326	−0.377	−0.429	−0.484	−0.540	−0.598	−0.658
	(1.77)	(1.59)	(1.43)	(1.28)	(1.15)	(1.03)	(0.92)	(0.83)	(0.74)	(0.67)
B8	−0.528	−0.593	−0.659	−0.726	−0.794	−0.862	−0.931	−1.000	−1.070	−1.140
	(1.21)	(1.11)	(1.01)	(0.92)	(0.85)	(0.78)	(0.72)	(0.67)	(0.62)	(0.57)
B9	0.000	0.000	0.000	0.000	0.000	0.000	0.000	0.000	0.000	0.000
	(2.00)	(2.00)	(2.00)	(2.00)	(2.00)	(2.00)	(2.00)	(2.00)	(2.00)	(2.00)
M1	2.152	2.233	2.322	2.416	2.515	2.617	2.723	2.831	2.941	3.053
	(0.05)	(0.06)	(0.06)	(0.06)	(0.07)	(0.06)	(0.06)	(0.06)	(0.06)	(0.05)
M2	2.498	2.593	2.692	2.795	2.900	3.008	3.118	3.229	3.342	3.455
	(0.01)	(0.01)	(0.01)	(0.01)	(0.02)	(0.02)	(0.02)	(0.01)	(0.01)	(0.01)
M3	2.775	2.883	2.993	3.105	3.218	3.332	3.448	3.564	3.682	3.799
	(0.00)	(0.00)	(0.00)	(0.00)	(0.00)	(0.00)	(0.00)	(0.00)	(0.00)	(0.00)
M4	2.192	2.233	2.278	2.326	2.377	2.429	2.484	2.540	2.598	2.658
	(0.01)	(0.02)	(0.02)	(0.03)	(0.03)	(0.03)	(0.03)	(0.04)	(0.04)	(0.04)

Line frequencies and relative intensities for the A_4B spin system

Line \ $\frac{J_{AB}}{(\nu_A-\nu_B)}$	0.95	1.00	2.00	3.00	4.00	5.00	10.00	∞
A1	2.661	2.781	5.236	7.723	10.217	12.713	25.207	—
	(0.22)	(0.20)	(0.04)	(0.02)	(0.01)	(0.01)	(0.00)	
A2	1.219	1.219	1.213	1.210	1.207	1.206	1.203	1.200
	(3.17)	(3.15)	(3.03)	(3.01)	(3.01)	(3.00)	(3.00)	(3.00)
A3	1.185	1.186	1.196	1.198	1.199	1.199	1.200	1.200
	(4.15)	(4.14)	(4.04)	(4.02)	(4.01)	(4.01)	(4.00)	(4.00)
A4	1.163	1.165	1.183	1.189	1.192	1.193	1.197	1.200
	(3.08)	(3.08)	(3.02)	(3.01)	(3.01)	(3.00)	(3.00)	(3.00)
A5	0.781	0.781	0.787	0.790	0.793	0.794	0.797	0.800
	(7.73)	(7.76)	(7.94)	(7.98)	(7.99)	(7.99)	(8.00)	(8.00)
A6	0.815	0.814	0.804	0.802	0.801	0.801	0.800	0.800
	(8.77)	(8.79)	(8.94)	(8.97)	(8.99)	(8.99)	(9.00)	(9.00)
A7	0.837	0.835	0.817	0.811	0.808	0.807	0.803	0.800
	(7.83)	(7.85)	(7.96)	(7.98)	(7.99)	(7.99)	(8.00)	(8.00)
A8	0.853	0.851	0.828	0.820	0.815	0.812	0.806	0.800
	(4.91)	(4.92)	(4.97)	(4.99)	(4.99)	(5.00)	(5.00)	(5.00)
A9	1.934	2.000	3.414	4.886	6.372	7.864	15.348	—
	(1.09)	(1.00)	(0.26)	(0.11)	(0.06)	(0.04)	(0.01)	
A10	1.276	1.281	1.318	1.326	1.329	1.331	1.333	1.333
	(3.49)	(3.46)	(3.14)	(3.06)	(3.04)	(3.02)	(3.01)	(3.00)
A11	0.724	0.719	0.682	0.674	0.671	0.669	0.667	0.667
	(10.86)	(10.96)	(11.71)	(11.87)	(11.93)	(11.95)	(11.99)	(12.00)
A12	0.785	0.781	0.732	0.712	0.702	0.695	0.681	0.667
	(8.47)	(8.50)	(8.83)	(8.92)	(8.95)	(8.97)	(8.99)	(9.00)
B1	0.714	0.719	0.764	0.777	0.783	0.787	0.793	0.800
	(4.78)	(4.80)	(4.96)	(4.98)	(4.99)	(4.99)	(5.00)	(5.00)
B2	−1.166	−1.281	−3.686	−6.156	−8.641	−11.133	−23.616	—
	(0.05)	(0.04)	(0.01)	(0.00)	(0.00)	(0.00)	(0.00)	
B3	−1.570	−1.686	−4.095	−6.564	−9.048	−11.538	−24.019	—
	(0.07)	(0.06)	(0.01)	(0.01)	(0.00)	(0.00)	(0.00)	
B4	−1.918	−2.037	−4.474	−6.951	−9.439	−11.931	−24.416	—
	(0.08)	(0.07)	(0.02)	(0.01)	(0.00)	(0.00)	(0.00)	
B5	−2.228	−2.351	−4.828	−7.320	−9.815	−12.312	−24.806	—
	(0.09)	(0.08)	(0.03)	(0.01)	(0.01)	(0.00)	(0.00)	
B6	0.491	0.500	0.586	0.614	0.628	0.636	0.652	0.667
	(7.91)	(8.00)	(8.74)	(8.89)	(8.94)	(8.96)	(8.99)	(9.00)
B7	−0.719	−0.781	−2.146	−3.598	−5.074	−6.559	−14.030	—
	(0.60)	(0.54)	(0.12)	(0.05)	(0.02)	(0.02)	(0.00)	
B8	−1.210	−1.281	−2.732	−4.212	−5.702	−7.195	−14.681	—
	(0.53)	(0.50)	(0.17)	(0.08)	(0.05)	(0.03)	(0.01)	
B9	0.000	0.000	0.000	0.000	0.000	0.000	0.000	0.000
	(2.00)	(2.00)	(2.00)	(2.00)	(2.00)	(2.00)	(2.00)	(2.00)
M1	3.166	3.281	5.686	8.156	10.641	13.133	25.616	—
	(0.05)	(0.05)	(0.02)	(0.01)	(0.00)	(0.00)	(0.00)	
M2	3.570	3.686	6.095	8.564	11.048	13.538	26.019	—
	(0.01)	(0.01)	(0.01)	(0.00)	(0.00)	(0.00)	(0.00)	
M3	3.918	4.037	6.474	8.951	11.439	13.931	26.416	—
	(0.00)	(0.00)	(0.00)	(0.00)	(0.00)	(0.00)	(0.00)	
M4	2.719	2.781	4.146	5.598	7.074	8.559	16.030	—
	(0.04)	(0.04)	(0.03)	(0.02)	(0.01)	(0.01)	(0.00)	

Line frequencies and relative intensities for the A_3B_2 spin system. Mixed transitions have been omitted

Line \ $\frac{J_{AB}}{(\nu_A-\nu_B)}$	0.00	0.05	0.10	0.15	0.20	0.25	0.30	0.35	0.40
A1	1.000	1.054	1.116	1.185	1.262	1.347	1.437	1.532	1.632
	(3.00)	(2.69)	(2.37)	(2.06)	(1.77)	(1.51)	(1.28)	(1.08)	(0.92)
A2	1.000	1.051	1.103	1.156	1.208	1.259	1.310	1.361	1.411
	(4.00)	(3.62)	(3.27)	(2.96)	(2.70)	(2.46)	(2.25)	(2.06)	(1.89)
A3	1.000	1.049	1.094	1.136	1.176	1.212	1.245	1.276	1.304
	(3.00)	(2.73)	(2.51)	(2.34)	(2.19)	(2.07)	(1.97)	(1.88)	(1.80)
A4	1.000	1.003	1.014	1.037	1.073	1.126	1.194	1.278	1.374
	(3.00)	(2.97)	(2.87)	(2.68)	(2.39)	(2.03)	(1.66)	(1.32)	(1.03)
A5	1.000	1.002	1.009	1.018	1.027	1.033	1.034	1.031	1.023
	(4.00)	(3.97)	(3.89)	(3.77)	(3.64)	(3.51)	(3.40)	(3.31)	(3.25)
A6	1.000	1.002	1.006	1.011	1.014	1.016	1.016	1.014	1.010
	(3.00)	(2.98)	(2.94)	(2.90)	(2.85)	(2.81)	(2.79)	(2.77)	(2.75)
A7	1.000	0.949	0.897	0.845	0.798	0.756	0.721	0.693	0.672
	(3.00)	(3.34)	(3.75)	(4.26)	(4.84)	(5.45)	(6.05)	(6.58)	(7.02)
A8	1.000	0.951	0.907	0.867	0.831	0.802	0.776	0.755	0.738
	(4.00)	(4.41)	(4.84)	(5.26)	(5.65)	(6.01)	(6.32)	(6.58)	(6.79)
A9	1.000	0.954	0.914	0.881	0.852	0.828	0.807	0.790	0.775
	(3.00)	(3.29)	(3.54)	(3.77)	(3.95)	(4.11)	(4.24)	(4.35)	(4.44)
A10	1.000	1.051	1.105	1.160	1.218	1.277	1.338	1.400	1.463
	(2.00)	(1.81)	(1.63)	(1.47)	(1.33)	(1.20)	(1.08)	(0.98)	(0.89)
A11	1.000	1.003	1.010	1.022	1.040	1.061	1.088	1.119	1.153
	(2.00)	(1.99)	(1.94)	(1.87)	(1.78)	(1.67)	(1.55)	(1.43)	(1.30)
A12	1.000	0.951	0.905	0.862	0.822	0.784	0.750	0.719	0.690
	(2.00)	(2.21)	(2.43)	(2.66)	(2.89)	(3.13)	(3.37)	(3.60)	(3.82)
A13	1.000	1.000	1.000	1.000	1.000	1.000	1.000	1.000	1.000
	(12.00)	(12.00)	(12.00)	(12.00)	(12.00)	(12.00)	(12.00)	(12.00)	(12.00)
B1	0.000	0.071	0.134	0.190	0.238	0.279	0.313	0.343	0.368
	(2.00)	(2.31)	(2.63)	(2.94)	(3.23)	(3.49)	(3.72)	(3.92)	(4.08)
B2	0.000	0.020	0.033	0.041	0.049	0.058	0.071	0.088	0.109
	(2.00)	(2.07)	(2.10)	(2.09)	(2.07)	(2.04)	(2.01)	(2.00)	(2.00)
B3	0.000	−0.030	−0.072	−0.125	−0.190	−0.265	−0.349	−0.441	−0.539
	(2.00)	(1.87)	(1.70)	(1.51)	(1.30)	(1.10)	(0.92)	(0.77)	(0.65)
B4	0.000	−0.075	−0.149	−0.222	−0.294	−0.365	−0.435	−0.504	−0.573
	(2.00)	(1.73)	(1.51)	(1.34)	(1.19)	(1.07)	(0.97)	(0.88)	(0.80)
B5	0.000	0.075	0.148	0.218	0.281	0.336	0.382	0.419	0.448
	(2.00)	(2.34)	(2.75)	(3.26)	(3.84)	(4.46)	(5.06)	(5.60)	(6.05)
B6	0.000	0.021	0.031	0.026	0.005	−0.034	−0.091	−0.165	−0.254
	(2.00)	(2.08)	(2.09)	(2.00)	(1.82)	(1.56)	(1.27)	(1.00)	(0.77)
B7	0.000	−0.028	−0.061	−0.096	−0.133	−0.169	−0.205	−0.242	−0.279
	(2.00)	(1.89)	(1.76)	(1.63)	(1.50)	(1.39)	(1.28)	(1.18)	(1.10)
B8	0.000	−0.079	−0.164	−0.256	−0.352	−0.453	−0.557	−0.665	−0.775
	(2.00)	(1.71)	(1.46)	(1.23)	(1.05)	(0.89)	(0.76)	(0.65)	(0.56)
B9	0.000	0.024	0.045	0.065	0.082	0.098	0.112	0.125	0.137
	(4.00)	(4.19)	(4.37)	(4.53)	(4.67)	(4.80)	(4.92)	(5.02)	(5.11)
B10	0.000	0.025	0.050	0.073	0.096	0.118	0.138	0.156	0.173
	(4.00)	(4.21)	(4.43)	(4.66)	(4.89)	(5.13)	(5.37)	(5.59)	(5.81)
B11	0.000	−0.025	−0.050	−0.073	−0.096	−0.118	−0.138	−0.156	−0.173
	(4.00)	(3.81)	(3.63)	(3.47)	(3.33)	(3.20)	(3.08)	(2.98)	(2.88)
B12	0.000	−0.026	−0.055	−0.087	−0.122	−0.159	−0.200	−0.244	−0.290
	(4.00)	(3.79)	(3.57)	(3.34)	(3.11)	(2.87)	(2.63)	(2.40)	(2.18)

Line frequencies and relative intensities for the A_3B_2 spin system. Mixed transitions have been omitted

Line \ $J_{AB}/(\nu_A - \nu_B)$	0.45	0.50	0.55	0.60	0.65	0.70	0.75	0.80	0.85
A1	1.736	1.843	1.952	2.064	2.177	2.292	2.408	2.525	2.643
	(0.78)	(0.67)	(0.58)	(0.50)	(0.44)	(0.39)	(0.34)	(0.30)	(0.27)
A2	1.462	1.513	1.565	1.618	1.672	1.726	1.782	1.839	1.897
	(1.74)	(1.61)	(1.48)	(1.37)	(1.26)	(1.17)	(1.08)	(1.00)	(0.93)
A3	1.331	1.355	1.377	1.398	1.417	1.435	1.451	1.466	1.479
	(1.73)	(1.67)	(1.62)	(1.57)	(1.53)	(1.49)	(1.45)	(1.42)	(1.39)
A4	1.480	1.593	1.712	1.835	1.961	2.089	2.218	2.348	2.478
	(0.81)	(0.64)	(0.51)	(0.41)	(0.34)	(0.29)	(0.24)	(0.21)	(0.18)
A5	1.013	1.000	0.986	0.972	0.958	0.944	0.931	0.919	0.907
	(3.21)	(3.20)	(3.20)	(3.22)	(3.24)	(3.27)	(3.30)	(3.34)	(3.37)
A6	1.005	1.000	0.994	0.988	0.981	0.975	0.969	0.962	0.956
	(2.75)	(2.74)	(2.74)	(2.75)	(2.75)	(2.76)	(2.77)	(2.77)	(2.78)
A7	0.657	0.645	0.636	0.630	0.625	0.621	0.618	0.615	0.613
	(7.38)	(7.66)	(7.88)	(8.05)	(8.19)	(8.30)	(8.39)	(8.46)	(8.52)
A8	0.723	0.711	0.701	0.692	0.685	0.678	0.672	0.668	0.663
	(6.97)	(7.11)	(7.23)	(7.33)	(7.41)	(7.48)	(7.54)	(7.59)	(7.63)
A9	0.761	0.750	0.740	0.731	0.723	0.716	0.710	0.704	0.699
	(4.51)	(4.57)	(4.62)	(4.67)	(4.70)	(4.73)	(4.76)	(4.78)	(4.80)
A10	1.528	1.593	1.659	1.726	1.794	1.862	1.931	2.000	2.070
	(0.81)	(0.74)	(0.67)	(0.62)	(0.57)	(0.52)	(0.48)	(0.44)	(0.41)
A11	1.192	1.233	1.278	1.326	1.377	1.429	1.484	1.540	1.598
	(1.18)	(1.06)	(0.95)	(0.85)	(0.77)	(0.68)	(0.61)	(0.55)	(0.49)
A12	0.664	0.640	0.619	0.600	0.583	0.567	0.553	0.540	0.529
	(4.02)	(4.21)	(4.39)	(4.55)	(4.69)	(4.82)	(4.93)	(5.03)	(5.12)
A13	1.000	1.000	1.000	1.000	1.000	1.000	1.000	1.000	1.000
	(12.00)	(12.00)	(12.00)	(12.00)	(12.00)	(12.00)	(12.00)	(12.00)	(12.00)
B1	0.389	0.407	0.423	0.436	0.448	0.458	0.467	0.475	0.483
	(4.22)	(4.33)	(4.42)	(4.50)	(4.56)	(4.61)	(4.66)	(4.70)	(4.73)
B2	0.132	0.157	0.182	0.207	0.232	0.255	0.277	0.298	0.318
	(2.01)	(2.03)	(2.06)	(2.10)	(2.14)	(2.19)	(2.23)	(2.28)	(2.32)
B3	−0.643	−0.750	−0.861	−0.974	−1.090	−1.207	−1.326	−1.446	−1.567
	(0.54)	(0.46)	(0.39)	(0.33)	(0.29)	(0.25)	(0.22)	(0.19)	(0.17)
B4	−0.642	−0.711	−0.780	−0.849	−0.918	−0.987	−1.056	−1.126	−1.195
	(0.74)	(0.68)	(0.62)	(0.58)	(0.53)	(0.50)	(0.46)	(0.43)	(0.40)
B5	0.470	0.487	0.500	0.511	0.520	0.527	0.533	0.538	0.542
	(6.41)	(6.69)	(6.91)	(7.08)	(7.21)	(7.32)	(7.41)	(7.48)	(7.54)
B6	−0.353	−0.461	−0.575	−0.694	−0.816	−0.941	−1.067	−1.195	−1.323
	(0.59)	(0.46)	(0.36)	(0.29)	(0.24)	(0.20)	(0.16)	(0.14)	(0.12)
B7	−0.317	−0.356	−0.397	−0.438	−0.482	−0.527	−0.574	−0.622	−0.672
	(1.01)	(0.94)	(0.87)	(0.80)	(0.74)	(0.68)	(0.63)	(0.58)	(0.53)
B8	−0.886	−1.000	−1.115	−1.231	−1.348	−1.466	−1.585	−1.704	−1.824
	(0.49)	(0.43)	(0.38)	(0.33)	(0.30)	(0.27)	(0.24)	(0.22)	(0.20)
B9	0.147	0.157	0.166	0.174	0.181	0.188	0.194	0.200	0.205
	(5.19)	(5.26)	(5.33)	(5.38)	(5.43)	(5.48)	(5.52)	(5.56)	(5.59)
B10	0.189	0.203	0.215	0.226	0.236	0.245	0.253	0.260	0.266
	(6.01)	(6.20)	(6.37)	(6.53)	(6.67)	(6.79)	(6.91)	(7.01)	(7.09)
B11	−0.189	−0.203	−0.215	−0.226	−0.236	−0.245	−0.253	−0.260	−0.266
	(2.80)	(2.73)	(2.66)	(2.60)	(2.55)	(2.50)	(2.46)	(2.42)	(2.39)
B12	−0.339	−0.390	−0.444	−0.500	−0.558	−0.617	−0.678	−0.740	−0.804
	(1.98)	(1.79)	(1.61)	(1.45)	(1.31)	(1.18)	(1.07)	(0.97)	(0.88)

Line frequencies and relative intensities for the A_3B_2 spin system. Mixed transitions have been omitted

Line \ $\frac{J_{AB}}{(\nu_A - \nu_B)}$	0.90	0.95	1.00	2.00	3.00	4.00	5.00	10.00	∞
A1	2.761	2.880	3.000	5.449	7.933	10.424	12.919	25.410	—
	(0.24)	(0.22)	(0.20)	(0.05)	(0.02)	(0.01)	(0.01)	(0.00)	
A2	1.957	2.017	2.078	3.423	4.868	6.341	7.826	15.295	—
	(0.86)	(0.80)	(0.74)	(0.22)	(0.09)	(0.05)	(0.03)	(0.01)	
A3	1.491	1.503	1.514	1.616	1.642	1.653	1.658	1.664	1.667
	(1.37)	(1.34)	(1.32)	(1.11)	(1.05)	(1.03)	(1.02)	(1.00)	(1.00)
A4	2.609	2.740	2.871	5.449	7.980	10.495	13.003	25.519	—
	(0.16)	(0.14)	(0.13)	(0.03)	(0.01)	(0.01)	(0.00)	(0.00)	
A5	0.896	0.886	0.877	0.782	0.757	0.747	0.742	0.736	0.733
	(3.41)	(3.44)	(3.47)	(3.82)	(3.92)	(3.95)	(3.97)	(3.99)	(4.00)
A6	0.950	0.944	0.939	0.863	0.827	0.806	0.793	0.764	0.733
	(2.79)	(2.80)	(2.81)	(2.91)	(2.95)	(2.97)	(2.98)	(2.99)	(3.00)
A7	0.612	0.610	0.609	0.602	0.601	0.600	0.600	0.600	0.600
	(8.57)	(8.62)	(8.65)	(8.91)	(8.96)	(8.98)	(8.99)	(9.00)	(9.00)
A8	0.659	0.656	0.653	0.625	0.617	0.612	0.610	0.605	0.600
	(7.67)	(7.70)	(7.73)	(7.93)	(7.97)	(7.98)	(7.99)	(8.00)	(8.00)
A9	0.694	0.690	0.686	0.646	0.631	0.623	0.619	0.610	0.600
	(4.82)	(4.84)	(4.85)	(4.96)	(4.98)	(4.99)	(4.99)	(5.00)	(5.00)
A10	2.140	2.210	2.281	3.732	5.212	6.702	8.195	15.681	—
	(0.38)	(0.36)	(0.33)	(0.11)	(0.06)	(0.03)	(0.02)	(0.01)	
A11	1.658	1.719	1.781	3.146	4.598	6.074	7.559	15.030	—
	(0.44)	(0.40)	(0.36)	(0.08)	(0.03)	(0.02)	(0.01)	(0.00)	
A12	0.508	0.509	0.500	0.414	0.386	0.372	0.364	0.348	0.333
	(5.20)	(5.27)	(5.33)	(5.83)	(5.93)	(5.96)	(5.97)	(5.99)	(6.00)
A13	1.000	1.000	1.000	1.000	1.000	1.000	1.000	1.000	1.000
	(12.00)	(12.00)	(12.00)	(12.00)	(12.00)	(12.00)	(12.00)	(12.00)	(12.00)
B1	0.489	0.495	0.500	0.551	0.567	0.576	0.581	0.590	0.600
	(4.76)	(4.78)	(4.80)	(4.95)	(4.98)	(4.99)	(4.99)	(5.00)	(5.00)
B2	0.337	0.354	0.371	0.551	0.615	0.646	0.664	0.700	0.733
	(2.36)	(2.40)	(2.44)	(2.83)	(2.92)	(2.96)	(2.97)	(2.99)	(3.00)
B3	−1.689	−1.811	−1.934	−4.428	−6.935	−9.441	−11.945	−24.455	—
	(0.15)	(0.14)	(0.12)	(0.03)	(0.01)	(0.01)	(0.00)	(0.00)	
B4	−1.265	−1.335	−1.406	−2.843	−4.312	−5.795	−7.284	−14.760	—
	(0.38)	(0.36)	(0.33)	(0.13)	(0.06)	(0.04)	(0.03)	(0.01)	
B5	0.546	0.549	0.552	0.577	0.584	0.588	0.591	0.595	0.600
	(7.59)	(7.64)	(7.67)	(7.92)	(7.97)	(7.98)	(7.99)	(8.00)	(8.00)
B6	−1.452	−1.581	−1.710	−4.271	−6.795	−9.306	−11.813	−24.324	—
	(0.10)	(0.09)	(0.08)	(0.02)	(0.01)	(0.00)	(0.00)	(0.00)	
B7	−0.723	−0.776	−0.830	−2.091	−3.497	−4.948	−6.419	−13.860	—
	(0.49)	(0.45)	(0.42)	(0.11)	(0.04)	(0.02)	(0.01)	(0.00)	
B8	−1.944	−2.065	−2.186	−4.646	−7.131	−9.623	−12.119	−24.610	—
	(0.18)	(0.16)	(0.15)	(0.04)	(0.02)	(0.01)	(0.01)	(0.00)	
B9	0.210	0.215	0.219	0.268	0.288	0.298	0.305	0.319	0.333
	(5.62)	(5.64)	(5.67)	(5.89)	(5.94)	(5.97)	(5.98)	(5.99)	(6.00)
B10	0.271	0.276	0.281	0.318	0.326	0.329	0.331	0.333	0.333
	(7.17)	(7.24)	(7.30)	(7.81)	(7.91)	(7.95)	(7.97)	(7.99)	(8.00)
B11	−0.271	−0.276	−0.281	−0.318	−0.326	−0.329	−0.331	−0.333	−0.333
	(2.36)	(2.33)	(2.30)	(2.09)	(2.04)	(2.02)	(2.02)	(2.00)	(2.00)
B12	−0.868	−0.934	−1.000	−2.414	−3.886	−5.372	−6.864	−14.348	—
	(0.80)	(0.73)	(0.67)	(0.17)	(0.07)	(0.04)	(0.03)	(0.01)	

References

1. Tiers, G. V. D.: J. Phys. Chem. **62**, 1151 (1958).
2. Hoffman, R. A., Gronowitz, S.: Arkiv Kemi **15**, 45 (1960).
3. Mooney, E. F.: An introduction to ^{19}F NMR spectroscopy. London: Heyden & Son Ltd. 1970.
4. Siddall, T. H., Stewart, W. E.: In: Progress in nuclear magnetic resonance spectroscopy, Vol. V (J. W. Emsley, J. Feeney, and L. H. Sutcliffe, ed.). Oxford: Pergamon Press 1969.
5. Stuart, R. D.: An introduction to Fourier analysis. London: Methuen 1961.
6. Ernst, R. R.: J. Chem. Phys. **45**, 3845 (1966).
7. Castellano, S., Bothner-By, A. A.: J. Chem. Phys. **47**, 5443 (1967).
8. Anderson, W. A., Freeman, R., Reilly, C. A.: J. Chem. Phys. **39**, 1518 (1963).
9. Lazlo, P.: In: Progress in nuclear magnetic resonance spectroscopy, Vol. III (J. W. Emsley, J. Feeney, and L. H. Sutcliffe, ed.). Oxford: Pergamon Press 1967.
10. Abraham, R. J., Bernstein, H. J.: Can. J. Chem. **39**, 216 (1961).
11. Corio, P. L.: Structure of high-resolution NMR spectra. New York: Academic Press 1966.
12. Dischler, B., Englert, G.: Z. Naturforsch. **16a**, 1180 (1961).
13. Swalen, J. D.: In: Progress in nuclear magnetic resonance spectroscopy, Vol. I (J. W. Emsley, J. Feeney, and L. H. Sutcliffe, ed.). Oxford: Pergamon Press 1966.
14. Hahn, E. L., Maxwell, D. E.: Phys. Rev. **84**, 1246 (1951).
15. Gutowsky, H. S., McCall, D. W., Slichter, C. P.: Phys. Rev. **84**, 589 (1951).
16. Bannerjee, N. K., Das, T. P., Saha, A. K.: Proc. Roy. Soc. (London), Ser. A**226**, 490 (1954).
17. Anderson, W. A.: Phys. Rev. **102**, 151 (1956).
18. Arnold, J. T.: Phys. Rev. **102**, 136 (1956).
19. Mc Connell, H. M., Mc Lean, A. D., Reilly, C. A.: J. Chem. Phys. **23**, 1152 (1955).
20. Pople, J. A., Schneider, W. G., Bernstein, H. J.: High-resolution nuclear magnetic resonance. New York: Mc Graw-Hill 1959.
21. Waugh, J. S., Dobbs, F. W.: J. Chem. Phys. **31**, 1235 (1959).
22. Whitman, D. R., Onsager, L., Saunders, M., Dubb, H. E.: J. Chem. Phys. **32**, 67 (1960).
23. Pople, J. A., Schaefer, T.: Mol. Phys. **3**, 547 (1960).
24. Diehl, P., Pople, J. A.: Mol. Phys. **3**, 557 (1960).
25. Hoffman, R. A., Gronowitz, S.: Arkiv Kemi **16**, 501 (1960).
26. Reilly, C. A., Swalen, J. D.: J. Chem. Phys. **32**, 1378 (1960).
27. Alexander, S.: J. Chem. Phys. **32**, 1700 (1960).
28. Bothner-By, A. A., Naar-Colin, C.: J. Am. Chem. Soc. **84**, 743 (1962).
29. — — Günther, H.: J. Am. Chem. Soc. **84**, 2748 (1962).
30. Castellano, S., Bothner-By, A. A.: J. Chem. Phys. **41**, 3863 (1964).
31. Hoffman, R. A.: J. Chem. Phys. **33**, 1256 (1960).
32. Diehl, P., Harris, R. K., Jones, R. G.: In: Progress in nuclear magnetic resonance spectroscopy, Vol. III (J. W. Emsley, J. Feeney, and L. H. Sutcliffe, ed.). Oxford: Pergamon Press 1967.
33. Corio, P. L.: Chem. Rev. **60**, 363 (1960).
34. Hoffman, R. A.: Pure Appl. Chem. **11**, 543 (1965).

MIX
Papier aus verantwortungsvollen Quellen
Paper from responsible sources
FSC® C105338
FSC
www.fsc.org

If you have any concerns about our products,
you can contact us on
ProductSafety@springernature.com

In case Publisher is established outside the EU,
the EU authorized representative is:
Springer Nature Customer Service Center GmbH
Europaplatz 3, 69115 Heidelberg, Germany

Printed by Libri Plureos GmbH
in Hamburg, Germany